Communications
in Computer and Information Science   1954

Rationale

The CCIS series is devoted to the publication of proceedings of computer science conferences. Its aim is to efficiently disseminate original research results in informatics in printed and electronic form. While the focus is on publication of peer-reviewed full papers presenting mature work, inclusion of reviewed short papers reporting on work in progress is welcome, too. Besides globally relevant meetings with internationally representative program committees guaranteeing a strict peer-reviewing and paper selection process, conferences run by societies or of high regional or national relevance are also considered for publication.

Topics

The topical scope of CCIS spans the entire spectrum of informatics ranging from foundational topics in the theory of computing to information and communications science and technology and a broad variety of interdisciplinary application fields.

Information for Volume Editors and Authors

Publication in CCIS is free of charge. No royalties are paid, however, we offer registered conference participants temporary free access to the online version of the conference proceedings on SpringerLink (http://link.springer.com) by means of an http referrer from the conference website and/or a number of complimentary printed copies, as specified in the official acceptance email of the event.

CCIS proceedings can be published in time for distribution at conferences or as postproceedings, and delivered in the form of printed books and/or electronically as USBs and/or e-content licenses for accessing proceedings at SpringerLink. Furthermore, CCIS proceedings are included in the CCIS electronic book series hosted in the SpringerLink digital library at http://link.springer.com/bookseries/7899. Conferences publishing in CCIS are allowed to use Online Conference Service (OCS) for managing the whole proceedings lifecycle (from submission and reviewing to preparing for publication) free of charge.

Publication process

The language of publication is exclusively English. Authors publishing in CCIS have to sign the Springer CCIS copyright transfer form, however, they are free to use their material published in CCIS for substantially changed, more elaborate subsequent publications elsewhere. For the preparation of the camera-ready papers/files, authors have to strictly adhere to the Springer CCIS Authors' Instructions and are strongly encouraged to use the CCIS LaTeX style files or templates.

Abstracting/Indexing

CCIS is abstracted/indexed in DBLP, Google Scholar, EI-Compendex, Mathematical Reviews, SCImago, Scopus. CCIS volumes are also submitted for the inclusion in ISI Proceedings.

How to start

To start the evaluation of your proposal for inclusion in the CCIS series, please send an e-mail to ccis@springer.com.

Prodipto Das · Shahin Ara Begum ·
Rajkumar Buyya
Editors

Advanced Computing, Machine Learning, Robotics and Internet Technologies

First International Conference, AMRIT 2023
Silchar, India, March 10–11, 2023
Revised Selected Papers, Part II

 Springer

Editors
Prodipto Das 🆔
Assam University
Silchar, India

Shahin Ara Begum 🆔
Assam University
Silchar, India

Rajkumar Buyya 🆔
The University of Melbourne
Melbourne, VIC, Australia

ISSN 1865-0929   ISSN 1865-0937 (electronic)
Communications in Computer and Information Science
ISBN 978-3-031-47220-6   ISBN 978-3-031-47221-3 (eBook)
https://doi.org/10.1007/978-3-031-47221-3

Preface

This book contains contributions of computer scientists, engineers, academicians and researchers from three different countries and various states of India, who have participated in the First International Conference on Advanced Computing, Machine Learning, Robotics and Internet Technologies (AMRIT 2023), held at Assam University, Silchar, Assam, India during 10–11 March 2023. The conference focused on four major thrust areas, *viz.* Advanced Computing, Machine Learning, Robotics and Internet Technologies.

Assam University, Silchar is a Central University established by an Act of the Indian Parliament (Act XXIII of 1989) and came into existence in 1994. Situated in the Barak Valley of southern Assam, the University is nestled in a sprawling 600-acre campus at Dargakona, about 23 km from Silchar town. In 2007, Assam University opened a 90-acre satellite campus at Diphu in the Karbi Anglong Hills District of Assam, thereby making quality higher education accessible to a wider section of society. With student strength about 5000 on both campuses, the University is a veritable melting pot of diverse communities, ideas and creativity. There are 43 postgraduate departments under 16 schools of study on the Silchar campus and 10 departments on the Diphu campus, offering a wide range of programmes geared towards equipping students and research scholars with knowledge, skills, experience and confidence. Apart from the two campuses of the University, there are 73 affiliated and permitted colleges in the five districts of south Assam, which together constitute the jurisdiction of Assam University. Besides, there are several Centres of Study in the University dedicated to research of high quality.

The Department of Computer Science was established in July 1997. The Department offers a 5-year integrated course leading to the Degree of M. Sc. in Computer Science. Provision exists for lateral exit and entry. The Department also offers a 2-year M. Sc. in Computer Science and a Ph. D. Programme. The thrust areas of the Department are Artificial Intelligence, Machine Learning, Natural Language Processing, Image Processing, Data Science, Soft Computing Techniques, Computer Networks and Security, Computer Architecture and Algorithms. The placement pattern of the passed-out students has been encouraging. Alumni of the Department of Computer Science of Assam University are now to be found in organisations such as IBM, TCS, Infosys, Oracle, HCL, WiproTech, Cognizant and many others including educational and research institutions in India and abroad.

AMRIT 2023 was the First International Conference in the Department of Computer Science. The conference aimed to gain significant interest among researchers, academicians, professionals and students in the region, covering topics of Computer Science and its current trends of research. AMRIT 2023 brought together researchers, educators, students, practitioners, technocrats and policymakers from academia, government, industry and non-governmental organizations and provided a platform for research collaboration, networking and presentation of recent research findings in the fields of Computer Science and allied subjects.

This event was a part of the Silver Jubilee Celebration of the Department. The Department has been providing a remarkable service to the region for the last 25 years and has reached a prestigious milestone with many achievements and remarkable success. The Department has successfully organised many national and regional programmes in the recent past on various research topics. The Department witnessed a grand National Conference, CTCS 2010, with funding from prestigious organisations like such as, DIT (now MeiTY), DoNER etc. In the recent past the research work in the Department in various thrust areas has touched a new level. The first International Conference AMRIT 2023 has initiated the journey to further heights. Five keynote addresses by reputed academicians, five high-quality invited talks by academicians and scientists from ISRO and BARC, one tutorial, technical sessions, poster presentations and industry-academia interactions were part of the conference. The participants and attendees of the conference came from Australia, the Philippines, Malaysia, Bangladesh and from different nationally reputed universities and organisations. Paper presenters, research scholars and student participants attend the conference both in physical and in virtual mode.

In the conference, a total of one hundred and ten papers were received, out of which forty-seven papers were accepted provisionally for this book volume. The review process was a single-blind peer review model. A total of 42 reviewers reviewed the papers with an average of 2–3 papers per reviewer. The reviewers were from various institutes and organisations with very high academic repute. The papers selected for the book volume were presented during the conference by the respective authors. The contents of the forty-seven papers are of high quality and cover new areas of Computer Science and allied subjects. The primary distinguishing features of this book are the latest novel research in the theme areas of the conference. This book contains systematic material for understanding the latest trends in Computer Science research and various challenging issues of Computer Science research and their solutions. We hope that this book volume will provide a valuable resource to the growing research community in the field of Computer Science.

The First International Conference on Advanced Computing, Machine Learning, Robotics and Internet Technologies (AMRIT 2023) at Assam University, Silchar was supported by SERB, Department of Science and Technology (DST), Govt. of India, Oil and Natural Gas Corporation, Silchar Asset, India and Assam University, Silchar, India.

March 2023

Prodipto Das
Shahin Ara Begum
Rajkumar Buyya

Organization

Conference Committee Members

Chief Patron

Rajive Mohan Pant   Assam University, India
   (Vice-chancellor)

Patron

Karabi Dutta Choudhury (Dean)   AESoPS, Assam University, India

General Chair

Shahin Ara Begum   Assam University, India

Technical Chair

Bipul Syam Purkayastha   Assam University, India

Publication Chair

Pankaj Kumar Deva Sarma   Assam University, India

Organising Chair

Prodipto Das   Assam University, India

Joint Organising Chairs

Saptarshi Paul   Assam University, India
Purnendu Das   Assam University, India
Biswa Ranjan Roy   Assam University, India
Debasish Roy   Assam University, India

Organising Members

Arindam Roy	Assam University, India
Rakesh Kumar	Assam University, India
Indrani Das	Assam University, India
Bhagwan Sahay Meena	Assam University, India
Rahul Kumar Chawda	Assam University, India

International Advisory Committee

Rajkumar Bhuyya	University of Melbourne, Australia
Neil P. Balba	Lyceum of the Philippines University – Laguna, The Philippines
Md. Fokray Hossain	Daffodil International University, Bangladesh
Mark Shen	National Central University, Taiwan
Is-Haka Mkwawa	University of Plymouth, UK

National Advisory Committee

Tanmoy Som	IIT-BHU, India
J. K. Mandal	Kalyani University, India
Utpal Roy	Visva-Bharati, India
Jamal Hussain	Mizoram University, India
N. P. Maity	Mizoram University, India
Rashmi Bhardwaj	Indraprastha University, India
Susmita Sur-Kolay	ISI Kolkata, India
M. K. Ghosh	ISRO, India
Shirshendu Das	IIT, Hyderabad, India
Samarjit Borah	SMIT, India
Ferdous Ahmed Barbhuiya	IIIT Guwahati, India
Nabanita Das	ISI Kolkata, India
Anajana Kakati Mahanta	Guwahati University, India
Smriti Kumar Sinha	Tezpur University, India

Technical Program Committee

Rajkumar Banoth	University of Texas at San Antonio, USA
Shayak Chakraborty	Florida State University, USA
Anjay Kumar Mishra	Madan Bhandari Memorial College, Nepal
Ricardo Saavedra	Azteca University, Mexico
S. Sharma	Texas A&M University Texarkana, USA
Goi Bok Min	Universiti Tunku Abdul Rahman, Malaysia

Megha Quamara	Institut de Recherche en Informatique de Toulouse, France
S. M. Aminul Haque	Daffodil International University, Bangladesh
R. Doshi	Azteca University, Mexico
M. Jahirul Islam	Shahjalal University of Science and Technology, Bangladesh
Rumel M. S. Rahman Pir	Leading University, Bangladesh
P. Shivakumara	University of Malaya, Malaysia
Koushik Guha	National Institute of Technology, Silchar, India
Wasim Arif	National Institute of Technology, Silchar, India
Sadhan Gope	National Institute of Technology, Agartala, India
Alok Chakraborty	National Institute of Technology, Meghalaya, India
Amitabha Nath	North Eastern Hill University, India
Abhishek Majumder	Tripura University, India
Bibhash Sen	National Institute of Technology, Durgapur, India
Shridhar Patnaik	BITS Mesra, India
Arnab Kumar Majhi	North Eastern Hill University, India
Buddhadeb Pradhan	UEM Kolkata, India
Santosh Satapathy	PDE University, India
Tapodhir Acherjee	Assam University, India
Abhijit Paul	Amity University, Kolkata & Assam University, India
Somen Debnath	Mizoram University, India
Sunita Sarkar	Assam University, India
Somnath Mukhopadhyay	Sikkim University, India
Arnab Paul	Assam University, India
Sourish Dhar	Assam University, India
Mousam Handique	Assam University, India
Subrata Sinha	Assam University, India
Abul Fujail	MHCM Science College, India
Rajib Das	Karimganj College, India
Ishita Chakraborty	Royal Global University, India
Munmi Gogoi	GLA University, India
Abhijit Paul	Gurucharan College, India
O. Mema Devi	Gurucharan College, India
Kh. Raju Singha	Nabin Chandra College, India
Munsifa Firdaus Khan	Assam Down Town University, India
Jhunu Debbarma	Tripura Institute of Technology, India
R. Chawngsangpuii	Mizoram University, India

Keynote Speakers

Rajkumar Buyya	University of Melbourne, Australia
Neil Perez Balba	Lyceum of the Philippines University – Laguna, The Philippines
Md. Forhad Rabbi	Shahjalal University of Science and Technology, Bangladesh
Susmita Sur-Kolay	Indian Statistical Institute, India
Prithwijit Guha	Indian Institute of Technology Guwahati, India

Invited Speakers

Mrinal Kanti Ghose	GLA University & ISRO, India
Debabrata Datta	Heritage Institute of Technology & Bhabha Atomic Research Centre, India
P. Shivakumar	University of Malaya, Malaysia
Utpal Roy	Visva-Bharati University, India
Prantosh Kumar Paul	Raiganj University, India

Neoteric Frontiers in Cloud, Edge, and Quantum Computing (Keynote)

Rajkumar Buyya[1,2]

[1] Cloud Computing and Distributed Systems (CLOUDS) Lab,
University of Melbourne, Australia
[2] Manjrasoft Pvt Ltd., Melbourne, Australia

Abstract. Computing is being transformed into a model consisting of services that are delivered in a manner similar to utilities such as water, electricity, gas, and telephony. In such a model, users access services based on their requirements without regard to where the services are hosted or how they are delivered. The cloud computing paradigm has turned this vision of "computing utilities" into a reality. It offers infrastructure, platform, and software as services, which are made available to consumers as subscription-oriented services. Cloud application platforms need to offer.

(1) APIs and tools for rapid creation of elastic applications and
(2) A runtime system for deployment of applications on geographically distributed Data Centre infrastructures (with Quantum computing nodes) in a seamless manner.

The Internet of Things (IoT) paradigm enables seamless integration of the cyber-and-physical worlds and opens opportunities for creating new classes of applications for domains such as smart cities, smart robotics, and smart healthcare. The emerging Fog/Edge computing paradigms support latency-sensitive/real-time IoT applications with a seamless integration of network-wide resources all the way from Edge to the Cloud.

This keynote presentation covers:

(a) 21st-century vision of computing and identifies various IT paradigms promising to deliver the vision of computing utilities;
(b) Innovative architecture for creating elastic Clouds integrating edge resources and managed Clouds;
(c) Aneka 5G, a Cloud Application Platform, for rapid development of Cloud/Big Data applications and their deployment on private/public Clouds with resource provisioning driven by SLAs;
(d) A novel FogBus software framework with Blockchain-based data-integrity management for facilitating end-to-end IoT-Fog/Edge-Cloud integration for execution of sensitive IoT applications;

(e) Experimental results on deploying Cloud and Big Data/IoT applications in engineering, health care (e.g., COVID-19), deep learning/Artificial intelligence (AI), satellite image processing, and natural language processing (mining COVID-19 research literature for new insights) on elastic Clouds;

(f) QFaaS: A Serverless Function-as-a-Service Framework for Quantum Computing; and

(g) Directions for delivering our 21st-century vision along with pathways for future research in Cloud and Edge/Fog computing.

Contents – Part II

Contents – Part I

Comparative Analysis of Different Machine Learning Based Techniques for Crop Recommendation

Rohit Kumar Kasera$^{(\boxtimes)}$ (iD), Deepak Yadav (iD), Vineet Kumar (iD), Aman Chaudhary (iD), and Tapodhir Acharjee (iD)

Triguna Sen School of Technology, Department of Computer Science and Engineering, Assam University, Silchar, Assam, India
rohitkumar.kasera@aus.ac.in

Abstract. Smart Agriculture is gradually becoming a blessing for mankind. It is very much efficient to mitigate food scarcity as well as minimize farmers' efforts to produce sufficient food. Crop recommendation is one area which is helping the farmers to choose the best crop for a particular soil and climate. This study compares various machine learning techniques, including "Random Forest" (RDMFR), "Decision Tree" (DCTR), K Nearest Neighbor (KNNB), Radial Basis Function Support Vector Machine (RBFSVMN) and Radial Basis Function Neural Network (RBFNUNT) for recommending crops. Our objective is to check that potato and onion crops are suitable for a given soil and climate or not. We had checked the type of soil depending on "Nitrogen" (N), "Phosphorus" (P), "Potassium" (K), temperature, rainfall and moisture sensor data. An assessment of the performance of an machine learning (MLRN) classifier model is conducted using hybrid K cross-validation. Comparative examination of the entire system has been performed with the previous existing system.

Keywords: Smart agriculture · NPK · Crop recommendation · RBFNUNT · DCTR · Stratified K-Fold

1 Introduction

Agricultural operations and farms today are very different from their predecessors decades ago due to technological advancements in the forms of sensors, gadgets, machinery, and information technology. Soil temperature, robotic and Soil dampness sensors, aerial imaging, high-frequency navigation system (GPS), and a variety of specialized internet of things (IoT) devices are just a few of the high-tech tools used in modern agriculture [1]. Businesses and farmers may operate more successfully, safely, financially, and sustainably with the aid of these cutting-edge agricultural technologies. The development of digital agriculture and the associated technology that underpin it has opened up several new knowledge opportunities [2, 3]. A complete farm can be connected with remote sensors, cameras, and other devices gathering data around-the-clock as mentioned in [4, 5]. These devices will monitor variables such as temperature, humidity,

P. Das et al. (Eds.): AMRIT 2023, CCIS 1954, pp. 1–13, 2024.
https://doi.org/10.1007/978-3-031-47221-3_1

and plant health [6, 7]. A vast amount of information will be produced by these sensors. Thanks to new technology farmers may get a lot better understanding of the state of things on the ground. A series of remote sensors gather information about the environment. This information is processed into algorithms and statistical data that farmers can use to manage their fields and make decisions accordingly. Crop recommendations in precision farming depend on a number of factors, including temperature, soil type, soil moisture, soil nutrients, rainfall, etc. MLRN models can recommend crops based on environmental, soil, and geographic factors. A difficult task is choosing a certain crop for a given season, place, and year. Although there have been a lot of earlier studies on this crop recommendation topic [8, 9], the systems still have some flaws in terms of making reliable predictions. Such as potato crops that depend on soil type and fluctuations in the value of soil nutrients is a great research field for optimizing crop recommendation process. One of the most recent studies on soil-based sensors for potato crops makes use of fertilizers to anticipate the outcome of the crop utilizing N, P, and K fertilizers [10]. The idea is for farmers to use these technologies to choose better fields in order to reach their goal of an improved crop. This analysis intends to improve the accuracy of the current crop recommendation system for potato crops based on climatic factors and soil nutrients. A traditionally, raw data are divided into training and testing is not the most effective way to evaluate model performance. Therefore, K-fold cross-validation is used in this paper as part of a hybrid methodology to train and test the dataset. Following is how the entire paper is structured: The problem statement and goal of this work are briefly introduced in Sect. 1, the existing research is described in Sect. 2, and the recommended methodology is covered in Sect. 3. In Sect. 4, an experimental outcomes and a comparison of various classifier methods are provided. Section 5 final discussion discusses the result.

2   Related Work

2.1   Crop Suggestions Based on Soil Series Using Machine Learning

Rahman et al. created a model or method that will enable us to classify the soil and foresee the crops that will thrive there [11]. Data on soil and crops are also used. A Soil class is determined by an MLRN-based classifier as "Support Vector Machine (SVM)".

2.2   Crop Recommendation System Based on Soil Analysis using Classification Method

Mariappan et al. created a model based on soil metrics, recommended fertilisers, and datasets that will enable to predict the best crop for a particular type of soil [12]. A list of suitable crops is provided by the proposed system, which maps soil and crop data. The "K nearest neighbour (KNNB)" based controller receives the various inputs from it.

2.3   Recommendation of Suitable Crops and Fertilizers Using Machine Learning

The "best-suited crops" for a certain region are suggested by the MLRN model, according to a paper by Jha. et al. [13]. The performance of different MLRN algorithms, such as "Naive Bayes (NVBS)", "Bayes Net (BN)", "Logistic Regression (LGRS)", "Multi-Layer Perceptron (MLPTR)", and RDMFR, is assessed based on both soil classification and crop recommendations for a given soil, using weighted "ROC", "True Positive", "False Positive", accuracy, and recall.

2.4   Machine Learning-based Crop Suggestions and Soil Classification

A technique for classifying soils according to their amounts of macro and micronutrients was proposed by Saranya, et al. [14]. Forecasting the kinds of crops that could be produced there was done using this approach. Four algorithms KNNB, Bagged Tree, SVM, and LGRS were used. Above all the methods, SVM offers the soil's average accuracy.

2.5   IoT and Machine Learning-Based Crop Recommendation

An IoT-based approach for analysing soil parameters in this model has been presented by Gosai, et al. [15]. The data gathered by these sensors will be saved by the microcontroller, who will then analyse it using a variety of machine learning techniques, including RDMFR and SVM.

2.6   A Machine Learning-Based Intelligent System as AgroConsultant for Crop Recommendation

A suggestion system and improved yield prediction were proposed by Doshi, et al. [16]. Based on a variety of environmental and geographic factors, the model gives farmers information on crops. To assess the precision of soil categorization, three algorithms are used: KNNB, DCTR, and Neural Network (NUNT). Over another MLRN algorithm, the NUNT technique yields the results that are the most accurate.

2.7   Using Machine Learning, Categorize Soil to Determine Which Crops will Thrive There

A system based on image processing was proposed by Jangir, et al. [17], in which soil sample digital images were processed by manipulating a "convolutional neural network (CNN)" to identify the soil sample's appropriateness for a certain crop. The CNN's goal was to suggest acceptable crops for evaluating the appearance of the soil samples.

2.8   Using Machine Learning, a System for Recommending Crops that uses an Ensemble Model and Majority Voting

The precision agricultural model was proposed by Pudumalar et al. [18]. An ensemble recommendation system is employed utilising "Random Tree", "CHAID", "KNNB", and

NVBS as learners with a majority voting method. Bagging is the most accurate method for yield prediction since it has the lowest mean absolute error and error deviation among the techniques listed above.

Further research is being done on crop prediction by soil analysis utilizing different MLRN algorithms all year long [19–21] and reaching varied probabilities.

3 Methodology

The suggested strategy lessens the constraints of earlier work by exhibiting varied levels of crop prediction accuracy. The literature review in Sect. 2 makes it obvious that ML classifier algorithms are widely used in research to address the issue of crop recommendation through soil analysis. Nevertheless, there is room for improvement in the prior research' error rate, accuracy, precision, and recall score. During preprocessing, a large dataset may have contained numerous samples that were repeated; this circumstance did not automatically disappear after training the dataset. When the output is inaccurate despite the model scoring 100%, there is a problem. As a result, the model is overfitted by the higher accuracy score. This can be called as overfitting situation [22, 23]. The recent work has been done to address the issue of overfitting for different dataset issues [24–26]. As a solution to the current issue, the proposed methodology uses "K-fold" "Stratified K-Fold" cross-validation [27, 28]. For recommendation of suitable crops, the experiment has been done with two crops potato and onion. The structure of the "K-fold" cross-validation" is shown in the Fig. 1.

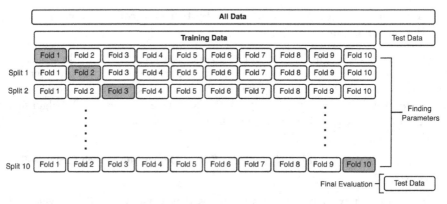

Fig. 1. Grid Search Architecture of 10 K-fold Cross Validation

The grid search architecture for the 10 K-fold cross-validations in Fig. 1 shows how data will be split into training and testing by applying 10 folds. Where k smaller clusters (k-1) are created by halving the training set. The effectiveness of the "Stratified K-Fold" is measured using the classical mean square error (CMSE) given in Eq. (1).

$$\text{CMSE} = \left(\frac{1}{n}\right) * \sum (y_i - f(x_i))^2 \qquad (1)$$

In the Eq. (1), n is the complete count of findings, y_i is the response weight of ith finding and $f(x_i)$ is the predicted weight of ith finding. K equal-sized groups is randomly selected from a dataset, in which one group is selected as holdout cluster and remaining K-1 cluster is fitted in the model. Calculation of overall test MSE is evaluated using Eqs. (2).

$$\text{TMSE} = \left(\frac{1}{K}\right) * \sum CMSE_i \qquad (2)$$

In the Eqs. (2), K is the number of folds, $CMSE_i$ is the TMSE on i^{th} steps.

The dataset has been collected from an online repository prepared by "AI Predictor App" [29]. The Table 1 indicates the dataset of the crop recommendation.

Table 1. Dataset crop recommendation

N	P	K	Temperature	Rainfall	PH	Crops
23.63	72.3	27.32	20.65	68.45	6.39	Potato
27.98	79.6	23.11	15.66	56.05	5.20	Potato
19.02	42.4	19.13	22.32	82.17	6.40	Onion
18.17	40.1	21.30	14.39	77.91	6.58	Onion

The outliers, null values, and negative values in the dataset were first eliminated in the preprocessing module. Afterwards a feature value and label value have been set following a features extraction. Potato and onion are the label value in this instance. The target labels are encoded with a value between 0 and n (classes-1) using the label encoder [30]. Where "x" is the feature set and "y" is the target variable. Then the dataset has been divided into training and testing groups with a size-to-ratio of 7:30. Then, for each sample data "x", the data for the training variable and testing variable are standardized using the standard scaler [31]. Using Eq. (3), the standard scaler score of "x" is determined.

$$scaler = \frac{(x - m)}{std} \qquad (3)$$

In the Eq. (3), x is the data sample, m is the training sample's average and std is the normative variation. Figure 2 shows the flow diagram for K-fold cross-validation.

In the Fig. 2 each MLRN classifier algorithm such as RDMFR, DCTR, NVBS, KNNB, SVM, RBFSVMN, RBFNUNT, Quadratic Discriminant Analysis (QDA), LGRS, and polynomial SVM (POSVM) cross-validation are carried out using K-fold "Stratified K-Fold". The dataset parameter was cross-validated using K-fold validation, where K is how many times the dataset parameter would be folded. K-1 of the folds are used to train the MLRN model on the data. On the remaining portion of the data, the MLRN findings are verified. It is used as a test set of data to gauge how well the model performs.

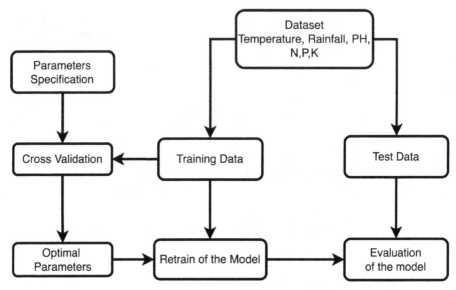

Fig. 2. Crop recommendation dataset splitting training and test with K- fold cross validation data flow diagram

4   Experimental Results and Analysis

The overall experimental setup has been conducted on "Google Colaboratory Notebook" using Python 3.10. The "scikit-learn = 1.2.1" package has been installed using the command "pip install scikit-learn = = 1.2.1" for importing the various libraries of the MLRN classification model for manipulation [32]. After installing the "scikit-learn" package and importing all its necessary libraries such as ("accuracy_score", "roc_auc_score", "classification_report", "cross_validate" etc.), the dataset has been loaded and converted to a ".csv" to extract the data into a "Pandas data frame". Then recognize the datatype, expected feature set, missing value, and target value. Data rows that have missing values are excluded during data cleaning. During the preprocessing of the data the label value has been specified as [1 for Potato and 0 for onion] using the string encoder library as "LabelEncoder", and the feature input variables has been specified as [Temperature, Humidity, Rainfall, PH, N, P, and K]. During data preprocessing, the sample features relationship is measured between each other. This includes temperature, rainfall, PH, N, P, and K using the "Seaborn matplotlib" "correlation heatmap" library for visualization. This experimental environment can also be set up on a local computer machine using the same Python configuration and the "Jupyter Notebook" platform. Table 1 displays the dataset of crop recommendation. The dataset contains a total of 13917 samples rows. The correlation visualization among all the sample feature variables for the class of potato and onion crops is shown in Fig. 3.

In the Fig. 3 correlations between features of a sample variables are used to determine how strong the relationship is between numeric values in the sample. Then feature set variables can be extracted based on this relationship. Various MLRN classifier methods

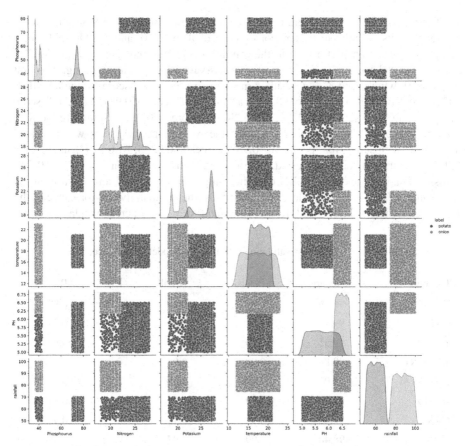

Fig. 3. Sample features data values comparison for onion and potato

(such as DCTR, RDMFR, RBFNUNT, KNNB, etc.) are inserted to the list for cross-validation and training data. In Table 2, the training parameter for the model is shown.

Table 2. Experimental parameter for various classifier model

Classifier Model	Experimental Parameter
RBFSVMN	Gamma = 0.3, Penalty error term (C) = 1.0, Kernel = rbf
DCTR	"max_depth" = 2
RDMFR	"max_depth" = 2, "n_estimators" = 10, "max_features" = 6
RBFNUNT	"num_of_classes" = 2, k = 100
Linear SVM	Kernel = "linear", C = 0.025
POSVM	Kernel = "poly", Degree = 3, Gamma = "auto"
KNNB	"n_neighbors" = 3

The rest of the classifier parameters (LGRS, NVBS, Bernoulli NVBS, and QDA) are set with its default parameter. For cv (K) equal to 7, 10, and 12, "K fold cross-validation" has been performed on the training data. This indicates that the MLRN classifier model has undergone numerous cross-validations during training. The first experiment, which used K = 7, resulted in an average accuracy score for the RDMFR, DCTR and RBFNUNT models of 98.35%, a "precision" score of 0.9815, a "recall" score of 0.9723, an "F1" score of 0.9783, and a "ROC" score of 0.9876. Similar to the K = 10 RDMFR, DCTR, and RBFNUNT models, the average accuracy score was evaluated as 98.93%, with scores for "precision" 0.9825, "recall" 0.9773, F1 0.9793, and "ROC" 0.9915. The training data was tested using K = 12 "cross-validation" after K = 10 fold "cross-validation", and the MLRN classifier outperformed K = 7 and 10 in terms of performance. The Table 3, Table 4 and Table 5 list the comparative performance of the MLRN classifiers at K = 7-fold, K = 10-fold, and K = 12-fold cross-validations.

Table 3. K = 7-fold cross validation classification model performance report

Classifier	Accuracy	Precision	Recall	F1	ROC
RBFSVMN	99.59	0.99	0.99	0.99	0.99
DCTR	99.48	0.99	0.99	0.99	0.99
RDMFR	99.51	0.99	0.99	0.99	0.99
RBFNUNT	98.87	0.99	0.98	0.99	0.99
Linear SVM	99.09	0.99	0.98	0.99	0.99
POSVM	99.40	0.99	0.99	0.99	0.99
LGRS	99.25	0.99	0.99	0.99	0.99
NVBS	98.48	0.99	0.97	0.98	0.98
Bernoulli NVBS	98.15	0.99	0.96	0.98	0.98
QDA	98.48	0.99	0.97	0.98	0.98
KNNB	99.34	0.99	0.98	0.99	0.99

The visualization graph of the various MLRN classification comparative analysis report at K = 12-fold is shown in Fig. 4. The y-axis represents the performance rate of the classifier algorithms for the different classification reports parameters, including accuracy, recall, precision, F1, and ROC score. The x-axis represents the list of classifier methods. Table 5 and Fig. 4 shows that, after K 12-fold cross-validation, the MLRN model RBFNUNT with the highest score 99.87 outperform the other MLRN models. The average accuracy rate of the above classification models was determined to be 99.18% using K12-fold cross-validation.

From the above experiment and analysis, applied classifier models are overly similar to each other. But the working principle of this model is different. As discussed in Sects. 2 and 3 several classifiers have been used to the crop recommendation system, but still have shortcomings based on model performance and overfitting. As an example, in the case of the RDMFR classifier, it is used to merge numerous models. This means it

Table 4. K = 10-fold cross validation classification model performance report

Classifier	Accuracy	Precision	Recall	F1	ROC
RBFSVMN	99.55	0.99	0.99	0.99	0.99
DCTR	99.50	0.99	0.99	0.99	0.99
RDMFR	99.35	0.99	0.98	0.99	0.99
RBFNUNT	98.67	0.99	0.97	0.98	0.99
Linear SVM	99.18	0.99	0.98	0.99	0.99
POSVM	99.41	0.99	0.99	0.99	0.99
LGRS	99.20	0.99	0.99	0.99	0.99
NVBS	98.21	0.99	0.96	0.98	0.98
Bernoulli NVBS	98.15	0.99	0.96	0.98	0.98
QDA	98.53	0.99	0.97	0.98	0.98
KNNB	99.45	0.99	0.98	0.99	0.99

Table 5. K = 12-fold cross validation classification model performance report

Classifier	Accuracy	Precision	Recall	F1	ROC
RBFSVMN	99.60	0.99	0.99	0.99	0.99
DCTR	99.50	0.99	0.99	0.99	0.99
RDMFR	99.53	0.99	0.99	0.99	0.99
RBFNUNT	**99.87**	0.99	0.99	0.99	0.99
Linear SVM	99.20	0.99	0.99	0.99	0.99
POSVM	99.44	0.99	0.99	0.99	0.99
LGRS	99.29	0.99	0.99	0.99	0.99
NVBS	98.46	0.99	0.99	0.99	0.98
Bernoulli NVBS	98.15	0.99	0.99	0.99	0.98
QDA	98.53	0.99	0.99	0.99	0.98
KNNB	99.46	0.99	0.99	0.99	0.99

creates a subset of DCTR and its outcomes are based on the majority ranking. However, DCTR recursively splits the data into subsets. So overfitting problems can arise in some algorithms, such as DCTR, RBFSVMN, LGRS, Linear SVM, QDA, and POVSVM. So applying "Stratified K-Fold" cross-validation techniques can reduce issues of recurrence of overfitting. The overall system comparative analysis has been discussed in Table 6.

Table 6 shows the results of a comparison between the proposed system and the previous existing systems, taking into account factors such as the Hybrid Validation Technique, Comparative Evaluation of Different MLRN Classifiers, Classification of

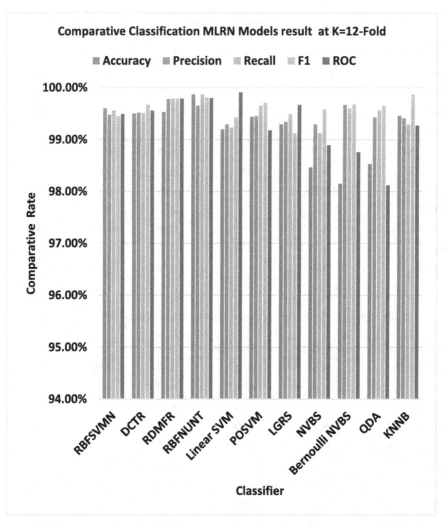

Fig. 4. Comparative classification report of MLRN at K12-fold cross validation

Soil NPK Values, Low Cost, and Optimized Prediction. By addressing the flaws in the earlier work, the proposed system improved results. Additionally, "Stratified K-Fold" cross-validation has been used to optimize the accuracy rate of the suggested system MLRN (RBFSVMN, DCTR, RDMFR, and RBFNUNT) models.

Table 6. Comparative analysis based on system performance with prior working system.

Reference	Hybrid validation technique	Comparative evaluation of different MLRN classifiers	Soil NPK values measured	Low cost	Optimized prediction
[10]	No	No	No	Yes	KNNB (90.00)
[13]	No	Yes	Yes	No	No
[15]	No	Yes	Yes	No	XGBoost (99.31)
Proposed System	Yes	Yes	Yes	Yes	RBFNUNT (99.87)

5 Conclusion

In the suggested methodology, a crop recommendation has been performed after applying multiple MLRN algorithms like DCTR, RDMFR, RBFSVMN and RBFNUNT to determine whether a given soil nutrient is acceptable for potato and onion crops. The average score from the MLRN model was 99.18%, which is higher than that of another existing system. The MLRN model dataset has undergone K-fold cross-validation with $K = 7$, 10, and 12 to combat the overfitting issue and assess correct prediction. Where the model performs better with $K = 12$ fold. The proposed system outperforms the total system comparison analysis with another earlier existing system. According to the findings and analysis it can be seen that the classical classification MLRN model performed better on the small-size dataset for crop recommendation, and RBFNUNT performance was better than the other existing classification models. In the crop recommendation problem due to a lack of understanding of weather variance, various crop types, and soil nutrients create challenges to classify which type of crop will grow in a particular geographical location with a specific time or season and RBFNUNT can face sensitivity related to the dimensionality of data and required the initial set of neurons and basic functions. So, using some hybrid methodology like RDMFR model with another neural network model can optimize the crop recommendation system. This can be developed in future work by utilizing IoT and MLRN, and predictions may be made based on real-time data acquired for the same potato, and onion crops including tomato.

References

1. Gaikwad, S.V., Vibhute, A.D., Kale, K.V., Mehrotra, S.C.: An innovative IoT based system for precision farming. Comput. Electron. Agric. **187**, 106291 (2021). https://doi.org/10.1016/j.compag.2021.106291
2. Jaiganesh, S., Gunaseelan, K., Ellappan, V.: IOT agriculture to improve food and farming technology. In: 2017 Conference on Emerging Devices and Smart Systems (ICEDSS), pp. 260–266. IEEE, Mallasamudram, Tiruchengode, India (2017). https://doi.org/10.1109/ICEDSS.2017.8073690

3. Lampridi, M., et al.: The cutting edge on advances in ICT systems in agriculture. In: The 13th EFITA International Conference. p. 46. MDPI (2022). https://doi.org/10.3390/engproc20210 09046

4. Reddy Maddikunta, P.K., et al.: Unmanned aerial vehicles in smart agriculture: applications, requirements, and challenges. IEEE Sensors J. **21**, 17608–17619 (2021). https://doi.org/10. 1109/JSEN.2021.3049471

5. Mekala, M.S., Viswanathan, P.: A novel technology for smart agriculture based on IoT with cloud computing. In: 2017 International Conference on I-SMAC (IoT in Social, Mobile, Analytics and Cloud) (I-SMAC), pp. 75–82. IEEE, Palladam, Tamilnadu, India (2017). https://doi.org/10.1109/I-SMAC.2017.8058280

6. Granwehr, A., Hofer, V.: Analysis on digital image processing for plant health monitoring. JCNS. **1**, 5–8 (2021). https://doi.org/10.53759/181X/JCNS202101002

7. Wang, S., Qi, P., Zhang, W., He, X.: Development and application of an intelligent plant protection monitoring system. Agronomy **12**, 1046 (2022). https://doi.org/10.3390/agrono my12051046

8. S, B.K., Parvathi, R.: Crop recommendation system by artificial neural network. Review (2021). https://doi.org/10.21203/rs.3.rs-874525/v1

9. Madhuri, J., Indiramma, M.: Artificial neural networks based integrated crop recommendation system using soil and climatic parameters. IJST. **14**, 1587–1597 (2021). https://doi.org/10. 17485/IJST/v14i19.64

10. Amkor, A., El Barbri, N.: Artificial intelligence methods for classification and prediction of potatoes harvested from fertilized soil based on a sensor array response. Sens. Actuators A **349**, 114106 (2023). https://doi.org/10.1016/j.sna.2022.114106

11. Rahman, S.A.Z., Chandra Mitra, K., Mohidul Islam, S.M.: Soil classification using machine learning methods and crop suggestion based on soil series. In: 2018 21st International Conference of Computer and Information Technology (ICCIT), pp. 1–4. IEEE, Dhaka, Bangladesh (2018). https://doi.org/10.1109/ICCITECHN.2018.8631943

12. Mariappan, A.K., Madhumitha, C., Nishitha, P., Nivedhitha, S.: Crop recommendation system through soil analysis using classification in machine learning. IJAST. **29**, 12738–12747 (2020)

13. Jha, G.K., Ranjan, P., Gaur, M.: A machine learning approach to recommend suitable crops and fertilizers for agriculture. In: Mohanty, S.N., Chatterjee, J.M., Jain, S., Elngar, A.A., Gupta, P. (eds.) Recommender System with Machine Learning and Artificial Intelligence, pp. 89–99. Wiley (2020). https://doi.org/10.1002/9781119711582.ch5

14. Saranya, N., Mythili, A.: Sri shakthi institute of engineering and technology: classification of soil and crop suggestion using machine learning techniques. IJERT **09**, IJERTV9IS020315 (2020). https://doi.org/10.17577/IJERTV9IS020315

15. Gosai, D., Raval, C., Nayak, R., Jayswal, H., Patel, A.: Crop recommendation system using machine learning. IJSRCSEIT **7**, 558–569 (2021). https://doi.org/10.32628/CSEIT2173129

16. Doshi, Z., Nadkarni, S., Agrawal, R., Shah, N.: AgroConsultant: intelligent crop recommendation system using machine learning algorithms. In: 2018 Fourth International Conference on Computing Communication Control and Automation (ICCUBEA), pp. 1–6. IEEE, Pune, India (2018). https://doi.org/10.1109/ICCUBEA.2018.8697349

17. Jangir, Y., Goyal, T., Kandari, S., Husain, A.: Soil classification and crop prediction using machine learning. In: Mehra, R., Meesad, P., Peddoju, S.K., Rai, D.S. (eds.) Computational Intelligence and Smart Communication, pp. 16–21. Springer Nature Switzerland, Cham (2022). https://doi.org/10.1007/978-3-031-22915-2_2

18. Pudumalar, S., Ramanujam, E., Rajashree, R.H., Kavya, C., Kiruthika, T., Nisha, J.: Crop recommendation system for precision agriculture. In: 2016 Eighth International Conference on Advanced Computing (ICoAC), pp. 32–36. IEEE, Chennai, India (2017). https://doi.org/ 10.1109/ICoAC.2017.7951740

19. Motwani, A., Patil, P., Nagaria, V., Verma, S., Ghane, S.: Soil analysis and crop recommendation using machine learning. In: 2022 International Conference for Advancement in Technology (ICONAT), pp. 1–7. IEEE, Goa, India (2022). https://doi.org/10.1109/ICONAT 53423.2022.9725901
20. Garanayak, M., Sahu, G., Mohanty, S.N., Jagadev, A.K.: Agricultural recommendation system for crops using different machine learning regression methods. Int. J. Agri. Environ. Inf. Syst. **12**, 1–20 (2021). https://doi.org/10.4018/IJAEIS.20210101.oa1
21. Chakraborty, A.P., Kumar, S.A., Pooniwala, O.R.: Intelligent crop recommendation system using machine learning. In: 2021 5th International Conference on Computing Methodologies and Communication (ICCMC), pp. 843–848. IEEE, Erode, India (2021). https://doi.org/10.1109/ICCMC51019.2021.9418375
22. Salam, M.A., Taher, A., Samy, M., Mohamed, K.: The effect of different dimensionality reduction techniques on machine learning overfitting problem. IJACSA **12**, (2021). https://doi.org/10.14569/IJACSA.2021.0120480
23. scikit, learn: Underfitting vs. Overfitting. https://scikit-learn.org/stable/auto_examples/model_selection/plot_underfitting_overfitting.html. Accessed 10 Feb 2023
24. Kahloot, K.M., Ekler, P.: Algorithmic splitting: a method for dataset preparation. IEEE Access. **9**, 125229–125237 (2021). https://doi.org/10.1109/ACCESS.2021.3110745
25. Joseph, V.R.: Optimal ratio for data splitting. Stat. Anal. **15**, 531–538 (2022). https://doi.org/10.1002/sam.11583
26. Joseph, V.R., Vakayil, A.: SPlit: an optimal method for data splitting. Technometrics **64**, 166–176 (2022). https://doi.org/10.1080/00401706.2021.1921037
27. scikit, learn: Cross-validation: evaluating estimator performance. https://scikit-learn.org/stable/modules/cross_validation.html. Accessed 10 Feb 2023
28. Xu, Y., Goodacre, R.: On splitting training and validation set: a comparative study of cross-validation, bootstrap and systematic sampling for estimating the generalization performance of supervised learning. J. Anal. Test. **2**, 249–262 (2018). https://doi.org/10.1007/s41664-018-0068-2
29. Chittupalli, S., Pande, S., Shah, T., Shirke, S., Shah, K.: AI-Based_Crop-Predictor_App. https://github.com/shan515/AI-Based_Crop-Predictor_App/tree/main/dataset. Accessed 10 Feb 2023
30. scikit, learn: sklearn.preprocessing.LabelEncoder. https://scikit-learn.org/stable/modules/generated/sklearn.preprocessing.LabelEncoder.html. Accessed 10 Feb 2023
31. scikit, learn: sklearn.preprocessing.StandardScaler. https://scikit-learn.org/stable/modules/generated/sklearn.preprocessing.StandardScaler.html. Accessed 2 Oct 2023
32. scikit, learn: Installing scikit-learn. https://scikit-learn.org/stable/install.html. Accessed 10 Feb 2023

Arduino Based Multipurpose Solar Powered Agricultural Robot Capable of Ploughing, Seeding, and Monitoring Plant Health

Hrituparna Paul[✉], Anubhav Pandey, and Saurabh Pandey

Department of Computer Science and Engineering, Shambhunath Institute of Engineering and Technology, Prayagraj, India
hrituoct@gmail.com

Abstract. Agribusiness is the primary occupation chosen by more than 40% of the worldwide populace. The types of equipment used to complete a variety of tasks in traditional farming operations are expensive and inadequately made to handle the tasks. So, in order to carry out farming operations, farmers require sophisticated equipment. The proposed study intends to construct a robot that can carry out tasks like grass cutting, spreading seeds and irrigating the plants. The proposed robot does not require an external power source because solar photovoltaic (pv) panels provide its power source. The PIC18F4520 interfaced Android application that controls the complete system transmits the signals to the robot for the necessary operations. Consequently, dc motors are used for seed planting and hard ploughing. For seed planting, consistent separation is maintained. Water is applied to the crop using a sprinkler with revolving nozzles. The grass-cutting tool consists of spinning blades with a knife edge sharpened on both sides to efficiently cut the excess grass. It combines multiple jobs, making it resourceful.

Keywords: agriculture · robot · ploughing · seed sowing · grass cutting · water sprinkling

1 Introduction

The majority of the rural population in India relies on agriculture as their primary source of income. The current farming practices include a lot of laborers and are manual or automated. In recent years, the amount of available labour has been consistently declining while earnings have increased. A better level of productivity is necessary. Therefore, it is necessary to create a device that aids farmers in solving the above- mentioned issue.

The Indian economy heavily relies on the agricultural sector. In India, agriculture is the main source of income. Since there are fewer farm labourers with advanced education, their demand is greater than ever and their pay is rising. Bullock carts, tractors, tillers, and other tools are typically used by humans to carry out farming tasks. The main issues in the agricultural sector are a shortage of manpower, a lack of understanding of soil testing, an increase in labour costs, seed and water waste, and a lack of available workers. It is relatively new to consider using robotics technology in agriculture. The potential is

P. Das et al. (Eds.): AMRIT 2023, CCIS 1954, pp. 14–24, 2024.
https://doi.org/10.1007/978-3-031-47221-3_2

to increase productivity using robots in agriculture is greater, and more and more robots are showing up on farms in various forms.

Many agricultural robots available today can only accomplish a single or a pair of jobs. By creating an agricultural robot that can spray water, plant seeds, mulch, and perform cutting operations, we are upgrading the robot. In the world, agriculture accounts for more than 42% of people's primary employment. Nowadays, agricultural industry has seen the emergence of intelligent and flexible autonomous vehicles.The21st century, is a century of creation, advancement, globalization, and so much other.

Other helpless health issues are the population's century, climate change, droughts, and cloud bursts. Many countries have developed automated agricultural robots systems to implement various agricultural outputs such as harvesting, monitoring, weeding, sowing, fertilizer and irrigation. Nonetheless, soil-based applications of seeding, fertilizing, and irrigation are included in this work. The goal of this work is to boost agricultural productivity while reducing the amount of labour required from farmers and speeding up their work. A robot that can operate both manually and automatically is helpful to people. In this project, the ARDUINO-controlled autonomous agriculture system's farm cultivation process is demonstrated.

Compared to conventional large tractors, mini autonomous machines are more efficient and human labour is being developed in the area of farm-based autonomous vehicles. These machines should be able to operate around-the-clock, throughout the year, in most weather situations, and with the intelligence necessary to act responsibly for extended periods of time in a semi-natural environment while performing the necessary duty. Autonomous vehicles have a variety of field operations that they can carry out that are more advantageous than using traditional machineries (Fig. 1).

Fig. 1. Manually working in field

This robot is intended to serve as a foundation for the creation of systems that will allow farming operations like the application of pesticides, fruit picking, and plant disease management to be automated. The system is made as modular as it can be, allowing for the creation or a change to any one of the separate tasks.

1.1 Spraying of Pesticides

Pesticides are required in agricultural fields to boost productivity, but they are also harmful to people and the environment. In the existing ways, farmers walk over the crop fields with a backpack sprayer that is manually handled by a human. They used to manually spray the pesticides in a precise manner.

Despite the use of certainty regarding pesticides equipment (separate head coverings and focused filtering systems for manual and automated spraying tactics, respectively), humans are nevertheless exposed to dangerous pesticides that can result in serious health issues.

Aside from health issues, there are disadvantages to both automated and manual spraying techniques. The motorized spraying doesn't have a clear aim and is designed to spray a harvest strip with rearranged stature (for example, the rancher may use shower spouts to spray only the grape bunches and not the natural product area). The lack of laborer's makes manual spraying monotonous, moderate, and limited.

1.2 Significance of Solar Energy

Solar energy is the vast amount of energy that the sun emits every day. More energy is emitted by it in a day than the entire planet needs in a year. This power originates within the sun itself. The Earth has blocked about 1.8/10 MW of solar energy, which is many times more than the rate at which energy is now being consumed on a worldwide scale. This has encouraged governments, researchers, and the power sector to increase their investments in the renewable energy sector with the goal of utilizing more of this clean energy and reducing global warming.

1.3 Mower

Agricultural and gardening workers have recently utilised conventional lawn cutters on a large scale. However, the energy used and the air pollution produced by the manually operated lawn cutters might have a negative impact on the wellbeing of the workers. As well as making a lot of noise and vibration, conventional lawn mowers do so, which can lead to major health problems like carpal tunnel, diminished finger blanching, white fingers, and hand sensitivity and dexterity.

Thus, we were able to combine these three tasks into a single robot by evaluating the afore mentioned three characteristics, which would boost farmer production and cut down on downtime. This inspired us to create a solar-powered model for pesticide application and lawn mowing. A four-wheeled, semi-automatic solar pesticide sprayer and mower powered by a wireless remote and a DC battery and made up of a solar panel, a battery, a motor, a pump, a container, cutting blades, and a microcontroller (Fig. 2).

The five sections that make up the paper are as follows: Sect. 1 contains the introduction, Sect. 2 differentiates between Existing System with Proposed System, Sect. 3 is the literature review of the latest research studies performed, Sect. 4 deals with the proposed model The Sect. 5 involves the experimentation and results and Sect. 6 proposes the conclusion and Future Scope.

Fig. 2. Modernized Technology

2 Existing System Vs Proposed System

The main issues in modern agriculture include a shortage of available agricultural labour, a lack of understanding of soil testing, a rise in labour costs, seed and water waste, and an increase in labour wages. The agricultural robot has been presented as a solution to all these drawbacks. The primary goal of an agricultural robot is to use robotic technology in the agriculture sector. The ploughing, sowing, and mud levelling tasks are efficiently carried out by the farm robot.

Robots are mechanical devices that can carry out a variety of functions without the need for human intervention. According to the controller's instructions, the robot operates. Along the robotic course, numerous parameters are sensed using a variety of sensors. As the brain of the robotic system, the microcontroller controls every single robotic system activity. By managing the DC motors, it also regulates wheel motion. The DC motors that drive the motion of the wheels, are driven by a motor driving circuit. A wireless Bluetooth connection between the robot and the smartphone allows for remote operation of the autonomous seeding robot for agricultural use. To control the robot, utilize the Bluetooth electronics app. Every single action the robot takes is managed by it.

3 Literature Survey

Kiran Kumar et al. in the paper [1] describes the design and development of a three-degrees-of-freedom, solar-powered, remote-operated pesticide spraying robot for agricultural usage. The prototype offered a respectable rate of area coverage at a respectably low operating cost.The degrees of freedom are provided through actuating devices that are quite sluggish and unreliable. Only pesticide spraying is permitted with this configuration.

Binod Poudel et al. in the paper [2] exemplifies the use of mechatronics and robotics in agriculture. The vehicle's resilience is not very good because it is a test model. Additionally, by fully eliminating human labour from this process, the safety and the farmers' long-term well-being are guaranteed. A petrol-based pesticide sprayer's performance is unaffected. The model can only be used for spraying pesticides because the system only has a two-wheel drive, which reduces its resilience. This technique fully eliminates labour. A petrol-powered pesticide sprayer's performance is unaffected.

Harshit Jain et al. in the paper [3] found that this particular pesticide sprayer with solar power is more affordable and produces reliable results when used for spraying. Because it uses solar energy, a non-traditional energy source, it is generally accessible and cost-free. As the system setup is not automated, moving it requires physical labour.

Vijaykumar N Chalwa et al. in the paper [4] gives an idea of the agricultural robot for a device that, after being improved for performance and cost, will show promise in agricultural spraying activities. Here Degrees of freedom do not exist for the Sprayer and totally reliant on power from the grid. It is too low of a setup height and is really expensive.

Vishnu Prakash K et al. in the paper [5] describes designing, utilising, and testing an autonomous multifunctional vehicle that operates safely, reliably, and economically. This autonomous vehicle traverses crop lines on agricultural property while performing duties that are hazardous or taxing for the farmers. Prior to any other configurations, the area is first prepared for spraying. However, additional configurations are also planned, such as a seeding system and a plug system to access the most visible portion of the plants and perform other tasks. (pruning, harvesting, etc.), and a vehicle to transport the fruits, crops, and agricultural waste. This robot's wheels are made to make moving across wet, squishy terrain simple.

Ankit Singh et al. [6] designs an agricultural robot that runs automatically. As one of the trends in 21st-century development on automation and cleverness of agricultural equipment, all types of agricultural robots have been studied and developed to apply to a significant increase in agriculture field in many countries. It performs basic tasks including picking, harvesting, weeding, pruning, planting, and grafting. To carry out numerous farm tasks like cutting and picking, they created a robot. The height of the crop and the location of the grass in the field are both determined via image processing. The mowed lawn and harvested crops are put in a container. The robot also has a device for spraying pesticides.

Mr.Sagar R. Chavan et al. in the paper [7] explains how Agriculture techniques have improved, such as the autonomous robot-assisted seeding of seedlings on tilled ground. A four-wheeled robotic vehicle with DC motor steering has been created. The vehicle is outfitted with a tool for planting seeds to evenly distribute the seed products. According to different seed products, the device will grow the plantation by taking specified rows and columns into account at a set distance. A sensor with an infrared wavelength takes into account and detects obstacles. A 12V battery pack with rechargeable batteries powers the entire assembly. By using solar electricity that is also attached to the robot, the battery pack may be recharged. This robot can prepare the ground, map out the positioning of the seeds, and reseed the area.

Nithin P V et al. in the paper [8] Designing, creating, and manufacturing an autonomous robot that can dig the ground, plant seeds, level the soil, and spray water are all integrated into larger systems that use solar energy and an electrical source. The robot is guided using a rack and pinion system. Motor power input is controlled by the relay switch. IR sensors are used to identify obstacles. The majority of people on earth—more than 40%—choose agriculture as their primary line of work, and recently, interest in the use of autonomous cars in agricultural has grown.

Ms. Aditi D. Kokate et al. in the paper [9] designed an automated machine that can do tasks including preprogrammed fertilisation, watering, and seeding. Moreover, both manual and automatic control are available. ARDUINO, which oversees the entire process, is the main component. Robots are currently replacing people in significant numbers in working duties, particularly in repetitive chores. The primary phase in farming is seeding. In this procedure, every row of the farming plot is seeded. The soil sensor is employed in the irrigation process to keep an eye on the weather. It monitors this level, notifies the farmer, and then gradually sprinkles a small amount of water on each row of seed that has been put in the farming plot. Yet, A lot of plants need fertilisers when the seed germinates and the seed begins to develop. The fertilization process is the same as the method of irrigation. Solar technology is used by the autonomous robot.

L. Manivannan et al. in the paper [10] proposes a firebird V automatic robot that aims to automate the entire agricultural process, with a focus on the production of onions. The robot known as Fire Bird V.

ATMEGA 8 is the slave controller, ATMEGA 2560 is the master controller, and other accessories include gripper designs and IR. The proposed system prototype is implemented by picking a location that takes into account any form of onion crop's agricultural field. The automatic robot selects the planting location using detectors and seed products that will be planted in the next field utilizing the gripper setup of the automatic robot.

4 Proposed Model

The primary goal of this work is to develop a multifunctional machine that can be used for a variety of tasks, including digging up soil, planting seeds, leveling muck, and spraying water, all while requiring the fewest accessory modifications and spending the least amount of money. Both battery and solar power are used throughout the robot's whole system.

This section contains the methodological process, the circuit diagram, and the block diagram. Hardware and software tools are integrated throughout the construction of the agricultural robot. The agricultural robot's block diagram is shown in Fig. 3.

The developed robot's master controller is a microprocessor called an Arduino Uno. Via Bluetooth communication, the robot's entire functioning is managed. The agricultural robot is a self-contained, remote-control robot that connects to a Smartphone through Bluetooth and can be wirelessly controlled from afar. Each and every action the robot takes is managed by the Bluetooth electronics app. The Bluetooth HC-05 module, which is permanently mounted on the robot, receives signals from the Bluetooth electronics app and transmits them to the microcontroller for processing.

- The robot's base frame has four connected wheels that are operated by dc motors in the back.
- A cultivator that is designed to dig the soil is mounted on one end of the frame and is also powered by a dc motor.
- To spray the water, use a water pump.
- The robot's top is covered with solar panels that are connected to the battery to charge the battery. As a result, both the solar panel and battery use the sun's energy to its fullest potential.
- The 12-V battery is needed to power the entire robot.
- This is operated via an IR transmitter and receiver.

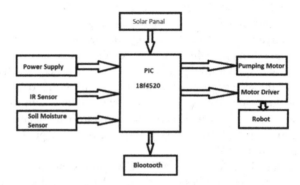

Fig. 3. The block diagram of multipurpose agriculture robot

5 Experimentation and Results

The power "ON" command starts it. The two sensors, an IR sensor and a moisture sensor, are turned "ON" when the robot moves ahead. These are employed for real-time environmental monitoring. In order to chop the undesired grasses or weeds, rotors are activated by IR sensors. Similar to this, water pumps are turned ON or OFF depending on whether the soil is damp or dry. In order to plough the ground, a robotic arm is built by switch (Fig. 4).

Only for control purposes, a robot is operated manually. The Robot is powered up and moved forward with this. The three key mechanisms are: (i) Rotor is turned ON/OFF through IR sensors identifying undesired grasses; (ii) Water pump is turned ON/OFF utilizing moisture sensor based on soil moisture detection. For plough the land, a switch is used to turn on and off robotic (iii) Automatic mode Robot operating status is monitored using an Android app with Bluetooth connectivity.

Steps involved in monitoring of Robot –

(1) The program looks for a robot with a Bluetooth interface (HC-05).
(2) The phone and the robot were once successfully connected.
(3) There are three different ways to communicate: byte stream mode, keyboard mode, and command line mode.

Fig. 4. Working of the Agricultural robot

(4) Choose the Byte stream mode for a linked device.
(5) In Byte stream mode, the agribot's operating condition must be one of the following: (i) Pumping motor ON; (ii) Rotor on for levelling or eliminating grass. Robot is going to do some excavating.

The solar panel used to power the robot uses sunshine to create electricity. The charging circuit receives this electrical energy in order to charge the battery to 12V. This battery powers the motor driver, controller, and other systems. The entire prototype of solar powered multifunction robot is shown in below (Fig. 5):

Fig. 5. Prototype of Robot

A variety of seeds, including wheat, rice, and gramme, were tested on typical agricultural soil with this prototype. We keep these seeds in a funnel. An Android application manages all robot functions using a Bluetooth device.

The robot begins to till the ground while concurrently dispensing seeds side by side when we give the commands "5" and "7." as shown in Fig. 6. For ploughing purposes, the plough is tilted downward direction at a particular angle to offer adequate dept for rows.

Using the command "8," the plough can be moved back to its original location. Funnels are mounted on a slider. Moving in the direction of command "5", the slider starts to move. The seeds used in this process were stacked in rows, but the spacing between them was uniform. The '6' command is used to stop the slider.. A circular cutter with sharp edges is part of the grass cutting machinery. Rotate the cutter by sending command 'a'. This cutter effectively removed weeds, as seen in Fig. 7. With the command "b," the cutter's rotation is stopped. A container is provided to hold the water. Send 'c' to sprinkle water on the field. Stop sprinkling by pressing "d".

A cone-shaped structure, hopper, or other container serves as the seed storage device, holding seeds for sowing. The hopper features a tube extension in its lower portion that will allow seed to flow to the robot's lower portion, or the seeding point. Through a seed dispenser, seeds are released from the hopper for planting. The seed dispenser assembly, which is attached to the motor's shaft, is made up of a motor and a tube with a hole in it.

Fig. 6. Seed Sawing and Ploughing Operation

The robot has three different types of sensors: one each for temperature, humidity, and soil moisture. To measure seed temperature, a temperature sensor is employed. The air's relative humidity is measured by the humidity sensor. The soil moisture sensors will gauge the amount of moisture present in the soil as well as its state, whether it is wet or dry. On the LCD panel, the output signals from various sensors are shown.

From the above results, we conclude the robot has performed ploughing, seed sowing, grass cutting, water sprinkling operations properly.

Fig. 7. Grass Cutting mechanism

6   Conclusion and Future Scope

Multipurpose farming robot has been successfully tested and actualized for tasks like plough, seed, mow, and sprinkle water. This investigation's main finding demonstrates that the majority of these frameworks that support self-government are more adaptable than conventional frameworks. Lessening human intervention, ensuring a proper water system, and making efficient use of resources are benefits of multipurpose horticulture robots. Small and medium sized farmers can use the spraying and mowing robot that has been developed. The spraying unit will be produced on a large scale, which will cut costs greatly and help Indian agriculture operations in several ways.

The seed can feed into the soil constantly and without any restrictions with the aid of cutting-edge seed-sowing machinery. The majority of the seed-sowing machinery mentioned above may be operated by a single person. Hence, it lowers labour costs. By using these seed-sowing tools, the overall cost of the seed-sowing procedure will be decreased.

In the future, we will use an Android application to construct a sensor-based monitoring system for soil moisture and temperature to improve user accessibility. Utilizing the Android App that will be made using App Inventor, a system for monitoring soil moisture, humidity, and temperature will be established. The app's interface will be made to show data on each land segment's soil moisture, humidity, and temperature.

References

1. Kumar Kiran, B.M., Rao Nagaraja, S., Pranupa, S.: Design and development of Three DoF Solar powered smart spraying agricultural robot. Test Eng. Manage. **83**, 5235–5242 (2020)
2. Binod, P., Ritesh, S., Ravi, S.B., Navaraj, S., Krishna Anantha, G.L.: Design and fabrication of solar powered semi-automatic pesticide sprayer. Int. Res. J. Eng. Technol. **04**(07), 2073–2077 (2017)

3. Harshit, J., Nikunj, G., Sumit, P., Harshal, G., Jishnu, G.: Designand fabrication of Solar pesticide sprayer. Adv. Res. Innov. Ideas (IJARIIE) **04**(2), 1715–1727 (2018)
4. Vijaykumar, C.N., Shilpa, G.S.: Mechatronics based remote controlled agricultural robot. Int. J. Emerg. Trends Eng. Res. **02**(7), 33–43 (2019)
5. Prakash Vishnu, K., Kumar Sathish, V., Venkatesh, P., Chandran, A.: Design and fabrication of multipurpose agricultural robot. Int. J. Adv. Sci. Eng. Res. **01**(01), 778–782 (2018)
6. Ankit, S., Abhishek, G., Akash, P.S.: Agribot: an agriculture robot. Int. J. Adv. Res. Comput. Commun. Eng. **04**(1), 62–64 (2015)
7. Sagar, C.R., Rahul, D., Shelke, Shrinivas, Z.R.: Enhanced agriculture robotic system. Int. J. Eng. Sci. Res. Technol. **01**(01), 368–371 (2015)
8. Nithin, V.P., Shivaprakash, S.: Multipurpose agricultural robot. Int. J. Eng. Res. **05**(06), 1129–1254 (2015)
9. Aditi, K.D., Priyanka, Y.D.: Multipurpose agricultural robot. Int. Adv. Res. J. Sci. Eng. Technol. **04**(02), 97–99 (2017)
10. Manivannan, L., Priyadharshini, S.M .: Agricultural robot. Int. J. Adv. Res. Electric. Electron. Instrument. **05**(01), 153–156 (2016)
11. Xia, C., Li, Y., Lee, J.: Vision based pest detection and automatic spray of greenhouse plant. In: IEEE International Symposium on Industrial Electronics (ISIE), pp. 920–925 (2009)

Galactic Simulation: Visual Perception
of Anisotropic Dark Matter

Anand Kushwah[1], Tushar Rajora[2], Divyansh Singh[3], Satwik Pandey[2],
and Eva Kaushik[3](✉)

[1] Electronics and Communication Engineering Department, Dr. Akhilesh Das Gupta Institute of
Technology & Management, New Delhi, India
[2] Computer Science and Engineering Department, Dr. Akhilesh Das Gupta Institute of
Technology & Management, New Delhi, India
[3] Information Technology and Engineering Department, Dr. Akhilesh Das Gupta Institute of
Technology & Management, New Delhi, India
kaushikeva0026@gmail.com

Abstract. To visualize the simulation of a galaxy. Since computation of billions
of particles is very rigorous, our main objective is to create an N-body simulation,
which will help us visualize over a hundred billion particles moving over time in
space, leading to the formation of a galaxy or clusters of stars and galaxies. This
simulation uses snapshots to explain the galactic simulation. Here each snapshot
possesses a unique identity over different properties like mass, location etc. This
paper would use a two-phase visualization method, one which bisects and collects
relevant data for the visualization and the second for projection on the arbitrary
plane of 3d points. In the first half of the application, programming is used to
host the cluster with the help of Single Program Multiple Data (SPMD) whereas
the second half focuses on the reduction of thousands of cores with the help of
GPU-accelerated computing and parallelization of matrix multiplication. With the
help of our application, we seek to speed up the computation of the output and the
distribution of these particles around the system's center of mass.

Keywords: N-Body Simulations · Dark Matter · Galaxies · Haloes · Cosmology
and Cluster

1 Introduction

Our aim is to envision a simulation of galaxy formation. Understanding how things
in the universe form, what were its initial conditions, and how the universe operates
and correlates with each other is a big task for researchers and astrophysicists to analyze and study since these things cannot be recreated in the laboratory. Various analytic
models are created to explore the universe's formation and understand its physical process to overcome this problem. Particularly, Simulations on dark matter help physicists
understand the structural growth of the universe more because dark matter shapes the
evolution of massive structures. In the present scenario, for computing cosmological

models of galaxy formation, there are two basic methods namely semi-analytic and self-consistent hydrodynamical simulations. Semi-analytic models give an explanation of baryonic physics by including galaxy formation as a post-processing step after the output of N-body simulations or on to Monte Carlo realizations of DM merger history trees [1]. This approach examines the baryonic physics that determine galaxy populations using straightforward analytical prescriptions. The models' free parameters are modified to accurately replicate some important findings, such as the local stellar mass function, and are then utilized to forecast additional features of the observed galaxy population. Furthermore, if these predictions to observation are altered, this model can be adjusted. While this method enables quick parameter space exploration of the fundamental physics at play, its predictive power for some observables is constrained. Based on the present theories, the structure formation leads on to the formation point that's characterized by mass in Kev's (Table 1).

Table 1. Hydrodynamical Structure Formation

Light Candidates	Mass in Kev
Standard Neutrinos	~1
Light Dilation	~0.5
Light Axino	~100
Majoron	~1
Mirror Neutrinos	~1

Self-consistent hydrodynamics is significantly more costly from a computational viewpoint than semi-analytic models. In this case, limitations are removed by linking the gas physics to the dark components and fully running hydrodynamical simulations that are completely self-consistent in cosmic volumes. This model's physics requirements are more extensive and difficult. Two factors contribute to the difficulties: (i) The hydro dynamical equations must be solved accurately and efficiently by the underlying numerical method, and it must have a sufficiently wide dynamic range to describe both cosmic and galactic scales and (ii) Physically and quantitatively significant implementations of the necessary baryonic processes are required. Here, We will be focusing more on shape, density law, lumpiness of the dark matter and its properties with respect to nearby galaxies. Dark matter is present in large amounts and constitutes a large amount of space [2]. There is a thought that dark matter is made of self-gravitating gasses which can be classified as cold (CDM) or hot (HDM) depending on the nature of the particles (Fig. 1).

HDM currently is incompatible with all theories around its structure and is not able to fulfill the criteria whereas CDM is researched using the help of a Newtonian framework.

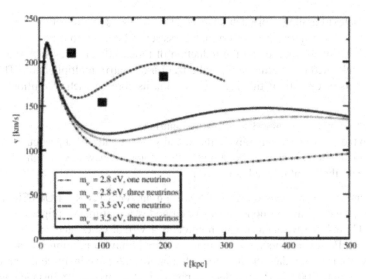

Fig. 1. Rotational Curves and Velocities far from the galaxy

2 Methodology

2.1 Working

To simulate a galaxy formation, a large amount of data of high resolution is used which thus produces a massive extensive amount of data. High Computational power is also required to analyze such data and produce output. High performance computing (HPC) refers to processing complex calculations at high speeds across multiple servers in parallel. Along with these clusters are composed of hundreds/thousands of compute servers that have been connected through a network. We aimed at the ricci service over the servers and after the creation of clusters the nodes have been added to the system. The N-body simulation we are visualizing comprises several hundred 3-dimensional ~1GB "snapshots" of 100 billion particles moving over time to form a galaxy. Each snapshot consists of an identification number, 3d location, mass, 3d velocity and potential energy of 100 billion dark matter particles. Analyzing and stimulating such data to produce results is a challenging task [3]. For this, we use measures to reduce the input data with our requirement to create the visualization.

We establish a two-phase visualization pipeline.

i) Data-cleaning/Down Sampling Phase-In this phase we select the data that seems to have relevant information from the massive pool of input simulation data.
ii) Projection/visualization phase - in this phase we perform a perspective projection onto an arbitrary viewing plane stake taking a cleaned and condensed dataset of 3d points.

We divide our application into two parts.

- In the first section, we use programming for the data we host on the cluster for which we use Single Program Multiple Data (SPMD). The parallelization is done using

SparkContext.Parallel() for each snapshot.This helps us in multi-threading, shared memory and distributed parallelisation across several cluster nodes.

- The Second phase focuses on the reduction of thousands of cores with the help of GPU-accelerated computing and parallelization of matrix multiplication. There are similar sets of operations that are performed at its core on each of millions of data points:-

    a. multiply by a particular 3×3 rotation matrix
    b. drop the z coordinate and divide the x, and y coordinates by a particular amount
    c. calculate which bucket in the output matrix holds the new x, y coordinates
    d. increase the count in that bucket by 1

Our First phase involves the use of Spark on an AWS EMR CLUSTER. For our application, we use 1 master node and 4 worker nodes as Optimal Configuration. Nodes have an EC2 m4.xlarge instance which contains 4 virtual CPUs and 16 GB of memory. Network Performance should be high and should have a minimum bandwidth of 750 Mbps. To accommodate the data to such a large extent, we increase the volume of EBS storage up to 100 Gb. For the Second phase, a g3.4xlarge AWS instance, with the Deep Learning Base AMI (Ubuntu, version 22.0) has been used. This instance accelerates code via a Tesla M60 graphics card, 2,048 parallel cores and 8 GB of GPU memory. The instance itself comprises 4 vCPUs and 35 GB of memory (Fig. 2).

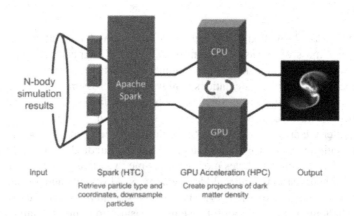

Input	Spark (HTC)	GPU Acceleration (HPC)	Output
	Retrieve particle type and coordinates, downsample particles	Create projections of dark matter density	

Fig. 2. Parallel Application

Running the application we need access to particle data of a N-body galaxy formation simulation. In particular, this application is tailor made for the Caterpillar simulation suite. This demands performing given sequence of operations on each of data points: 1) multiplication by a 3×3 rotation matrix 2) next drop the z coordinate and divide the x, y coordinates by a specific value 3) Identify the bucket in the output matrix that holds new x, y coordinates 4) increment the bucket count by 1 [4].

2.2 Phase-1

Since our primary focus is on particles of low mass, we eliminate particles of high mass i.e. type 2–5 particles since these particles are very few in number. The first pipeline section loads these files into Spark RDDs. We Redeem information particle types, the 3D positions, and mass from these files. We use the RDD.sample function to subsample the particles of type 1 [5]. This step ensures a more efficient administration of the pipeline. We finally join the x, y, and z positions of subsamples using the RDD function. Steps for building h5py:- For installing h5py, we built a bash script and added it as a bootstrap action when developing the cluster. 1. Create file with .sh extension. 2. Upload the file to amazon Simple cluster Service Bucket. 3. While uploading using advanced options configuration, add bootstrap actions. 4. To add bootstrap actions, select bootstrap action from Add bootstrap action dropdown menu and click configure and add. 5. You will notice a pop-up menu at the bottom. You can leave the default name or change the name accordingly. 6. Provide the location of your script from your S3 bucket by clicking on the folder icon. 7. Add additional arguments if necessary otherwise continue creating your cluster.

RUN THE FOLLOWING COMMAND, TO EXECUTE THE SPARK APPLICA-TION ON THE CLUSTER

spark-submit --num-executors 4 --executor-cores 4 project_spark.py Which collects the input data from folders in /mnt/s3 and generates outputs in the same folder. For different snapshots, simply change the 'snap' parameter in the code to get a different result. Our system OS and version is Amazon Linux AMI 2017.03. We are using Python 2.7.12 and Spark 2.2.0.

2.3 Phase-2 GPU Acceleration

To read 3.44 million points in 3d space and that too for each time period a .npy file Has to be visualized. For this to work, we developed code for both CPU and GPU. For a snapshot, we produce 40 different points of each individual snapshot by rotating the camera in a circle around the center of mass of a snapshot. Viewpoints are created when the datum of snapshots is:- 1. multiplied by a 3×3 matrix defined by the camera's position and angle 2. dropped if its resulting position is behind the camera 3. projected onto the viewing plane by dividing the new x and y coordinates by z and dropping z 4. assigned to a pixel in the output, and tallied there. Each different viewpoint is saved to disk as a png file and each snapshot is stored in a separate folder. Most Operations of the CPU code are performed using NumPy in lower-level lang to take advantage of multiple threads on certain operations.CuPy is primarily used in the case of the replacement of NumPy by GPU. Since CuPy was not implemented in aggregation operation, so Numba was used to operate its operation slightly at a lower level compared to NumPy.Hence, phase 2 depends on NumPy, CuPy, and Numba as well as a CUDA-enabled GPU with appropriate drivers [6]. We work from an AWS EC2 instance, selecting a g3.4xlarge instance for access to a GPU and the Deep Learning Base AMI (Ubuntu, Version 22.0) to handle the setup of NumPy, matplotlib, and the appropriate drivers.

2.4 Performance Evaluation

SPEED-UP, SCALING, OVERHEAD, AND OPTIMIZATION

A. Phase I: Spark

Following we plot a Graph which shows the speedup, as a function of the number of executors launched for the different numbers of nodes. This speedup is measured on the basis of our Spark code of a single snapshot. Here we see that 2 nodes and 4 cores per node have a factor of around 2.4 approximately. Similarly, with 4 nodes we get a speedup as high as 3.5 when we use 4 executors. But, we see a decline in the speedup when the number of nodes increased to 8 since we got a speedup of only 4.2 leading to much lower efficiency, or speedup per node when compared with 2 or 4 nodes. This is to be expected as there is a significant amount of sequential work in I/O and other overheads. The main reason for the decline of speedup with 4 nodes is that we invoke more than 4 executors, each virtual core is then split to run multiple threads which do not mainly improve performance. Hence we used 4 nodes and 4 executors for our goal to achieve a subtle speedup while maintaining speed-up efficiency [7]. The following Plot shows the speedup as a function of a number of executors running on a 4-node cluster with various numbers of executor cores per node. Here, maximum speedup decreases with an increase in the number of executor cores. The main reason for this is due to memory allocation constraints. Our application consumes >90% (11 GB out of 12 GB, as shown in the figure below) of memory in the first phase as it loads large data files onto memory for data reduction and particle downsampling. Since the first phase of our application is memory limited, it is not able to take advantage of the number of executors available.

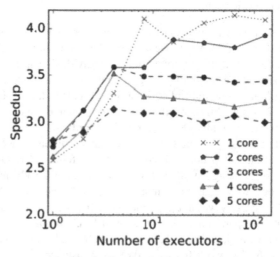

Fig. 3. Speedup by Executors

The above screenshot shows us the memory usage per node during runtime for a 4-node cluster using 1 executor core. Nine containers were allocated and distributed amongst the nodes. All nodes with two assigned containers during runtime used 11 GB of the 12 GB available memory, and the one node that was assigned 3 containers used 11.88 GB of the 12 GB available memory. These stats also tell us that there is high usage of memory. Is involved. We also note that on average only 2 VCores of the available 8 VCores are leveraged per node during runtime and this is uniform across all our tests regardless of whether we have 8 nodes or 2 nodes in a cluster. By Leveraging more than 90% of the available VCores we would have had a significant impact in speeding up our execution time when we increase the number of threads per core but this does not happen as Figs. 3 and 4 show we actually did not gain any performance because we only had at most 2 VCores available per node thereby decreasing our performance as the number of threads per core increased. Our main memory usage comprises its I/O, as the number of tasks increases, request for allocated memory also goes up similarly vice versa. Hence the data reduction phase of our application is banded with memory limitations.

B. **Phase II: GPU Acceleration**

This figure shows the total time spent processing 4 (of 255) snapshots The time needed to write output images to disk is roughly constant across all three environments, while the time spent on the actual computations is drastically reduced by running on GPU, with speedups in the tens, the twenties, or hundreds. Interestingly, the histogram operation completes much faster on the AWS instance than locally. The timings were repeated three times, and are remarkably consistent from run to run.

3   Background Related Work

Shohei Aoyama et al. [8] In recent decades, the significance of cosmic dust in astrophysical processes has gained recognition. Dust plays a crucial role as a highly efficient catalyst for the production of molecular hydrogen (H2) in the interstellar medium. Additionally, dust cooling regulates the typical mass of the final fragments in star-forming clouds. In the context of protoplanetary disks, the growth of dust particles ultimately facilitates the formation of planets. Moreover, dust particles actively participate in radiative processes by absorbing and re-emitting astral light in the far infrared, thereby altering the spectral energy distribution of galaxies. The size distribution of dust particles significantly influences the properties of galactic dust, particularly the extinction curve, which represents the wavelength dependence of absorption and scattering cross sections. To accurately estimate the star formation rate (SFR) of a galaxy, corrections for dust extinction are necessary. The particle size distribution, which determines the surface area, is directly linked to the rate of hydrogen molecule formation.

B.W. Keller et al. [9] the galaxy functions as a stellar production facility, gathering large amounts of gas within substantial potential wells and converting it into stars. The energy released by stars through winds and supernovae (SNe) serves as the controlling factor in this process. In the absence of a significant energy source for stirring and heating the interstellar medium (ISM), star formation depletes all available gas. Stellar feedback, which allows for self-regulation of star formation, is essential and one of the primary mechanisms responsible for generating a diverse ISM with multiple phases. The third aspect of feedback significantly influences this dynamic. It involves the redistribution (and even expulsion) of gas through outflows within the galaxy. Galactic winds, propelled beyond the escape velocity of the galaxy, can remove gas capable of forming stars from the galaxies. Gas fountains, on the other hand, circulate cold gas from the galactic disk through the high regions of the disk and the galactic halo, serving as temporary storage. These reservoirs in the galactic plane, too hot and diffuse to support star formation, are likely the ultimate fate of the galaxy and a plausible mechanism for the observed concentration of metals around the circum-galactic medium.

Eric J. Baxter et al. [10] existence of large amounts of dark matter. The galactic longitude scale is already an established fact. Currently, this dark matter is made of relique self-gravitating gases which are labeled as "cold" (CDM) or "hot" (HDM), depending on the relativistic or non-relativistic properties of particle energy spectra decoupling from the cosmic mix. HDM scenarios are not favored as they appear conflicting with current theories of structure formation. Normal CDM Particles examined in the Newtonian frame can be considered non-interacting (collision-free particles) or self-interacting. However, recent numerical simulations show that the non-interactive CDM model contradicts galactic-scale observations as follows: (a) 'substructure problems' associated with excessive clustering at sub-galactic scales. (b) "cusp problem" characterized by a monotonically increasing density in the direction of the center of the halo, the core is overly concentrated. To address these issues, the possibility of self-interacting dark matter was considered, thus producing non-zero pressure or thermal effects, a self-interactive model of CDM.

Steven R. Majewski et al. [11] in the last decade, cosmological observations have focused on normal matter (e.g., protons, neutrons, atoms, etc.) make up only one-fifth of the matter in the universe, the remaining 80% being Dark Matter, the identity of this Dark Matter is a mystery, but its properties influence the way galaxies form and the universe evolves. Its existence is also one of the strongest pieces of evidence of the fundamental incompleteness of the standard model of particle physics. Natural in that sense. The Dark Matter problem is one of the biggest problems in both astronomy and particle physics. The microscopic nature of Dark Matter affects how it accumulates around galaxies and can therefore be studied through astronomical observations Dynamic measurements, They provide important and unique insights into the properties of the Dark Matter in and around our Milky Way, which are particularly sensitive to how the Dark Matter is distributed on the galactic scale. In today's leading models of structure formation, Dark Matters are mostly 'cold', consisting of particles much more massive than protons and capable of agglomerating. Strong on a small scale. Cold-Dark Matter models of structure formation predict that the Milky Way must be surrounded by diffusely extended 'Dark Matter halos', which themselves are flooded with smaller self-connected Dark Matter clusters called 'sub-halos'. These sub-halos orbit the galaxy like satellites, and the satellite halo mass function is predicted to increase with decreasing mass. The minimum mass and total number of dark satellites are closely related to the graininess of the Dark Matter itself.

Luis G. Cabral-Rosetti et al. [11] a large body of evidence points to the existence of non-baryonic dark matter. There are three ways to directly check this hypothesis: 1. Produce dark matter or its cousins at accelerators, 2. Detect dark matter directly by the particles colliding with Earth in underground detectors, and 3. In space indirectly detects dark matter by observing the annihilation products of two dark matter particles. The current excitement in this space stems from the serendipitous maturation of all three of these technologies. The Large Hadron Collider was commissioned in 2009, and numerous direct detection experiments have demonstrated his ability to scale to the 1-ton level, with several experiments (Fermi Gamma-ray Satellite Telescope, Atmospheric Cherenkov Telescope, PAMELA, Ice cube) is available. Ready to recognize indirect signals

4 Application

This paper gave us a brief explanation of how far we can imagine and gaze back to the universe's beginning in order to discover additional new galaxies and universes. These simulation results show scientists and researchers what is feasible when it comes to the formation of galaxies and more innovative ways of what galaxies around the Milky Way result in, allowing us to more clearly comprehend what telescopes do for us. It would be possible to examine and study briefly the various physical processes involved in form-ing structures and galaxies in the Universe, composed of ordinary matter, dark energy, and dark matter. These processes include gravitation, gas cooling, star formation, super-nova feedback, supermassive black hole feedback, stellar evolution, radiation, magnetic fields, cosmic rays, and more [12]. Starting from smooth initial conditions constrained by observations of the cosmic microwave background, cosmological simulations enable detailed studies of the formation and evolution of structures and galaxies in the cosmos. These studies result in precise predictions of the galaxy population at various epochs of the Universe [13]. In order to assess and constrain such theories in the context of struc-ture and galaxy formation, it is additionally possible to investigate alternative forms of dark matter, dark energy, and gravity through suitable modified simulation methods. By comparing these results to observational data such as galaxy surveys, these studies pro-vide significant insights into the overall cosmological framework of structure formation and cosmological parameters.

5 Result

These pictures show the evolution of dark matter density over time as a Milky Way-mass galaxy forms in the center. The camera rotates around the center of mass of the system. Only the high resolution particles are shown. The simulation was designed such that the area directly around the central galaxy has only high resolution (low mass) particles, while lower resolution particles exist farther away from the galaxy. This is why in the visualization there are little to no particles near the edges of the frame; this area would have low resolution particles, but they are not of interest for our science goals. The 'z' value refers to the redshift at a given time (Fig. 5).

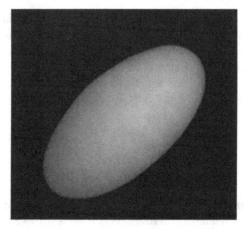

Fig. 4. Evolution of Dark Matter Density over Time

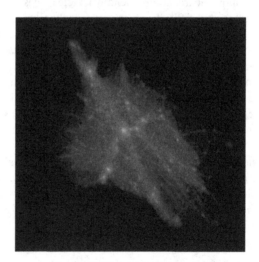

Fig. 5. Final Galaxy visualization

6   Conclusion

The origin of the universe remains a mystery, and it is constantly expanding, colliding, and experiencing supernovas, and other phenomena. Scientists have been attempting to model how the billions of galaxies in the observable universe formed from dust and gas after the big bang for many years. These simulations will assist us in our exploration, understanding of the workings of the universe, and discovery of the undiscovered. We created this simulator to aid in understanding the finding, which takes numerous large-scale images and chooses the best ones for the following stage. In the II phase, we take these pictures and display the findings from 40 angles. We aim to uncover additional unforeseen insights about the cosmos with the help of more improved technology in the future.

7 Future Scope

Since the creation of galactic simulators, visualizing the galaxy has become quite simple. We have achieved and learned how far we can stretch our limitations in understanding the formation of galaxies.

1. For future courses, our main aim would be to enhance data storage and access, since it is the most laborious section of the application. We were able to use EBS storage on a cluster to hold our data, but moving the data into the storage was a time-taking process. Furthermore, for larger, higher-resolution simulations, the EBS storage would likely be insufficient. Moreover, for the storage, we could look for other methods or focus on existing spark directly onto the primary storage cluster for the simulations.
2. Another area where we can bring modifications would be to include more visualization features in the application such as more color options, changes in angles or a more user-friendly interface.
3. Phase II could yet be improved. Lower-level "save" procedures might function more quickly than matplotlib. We could combine the output on a single disc if we divided the job across several computers. When handled on a single thread, this will produce quick computation, improving run time.
4. A better simulation could be obtained with the help of precise targeted users, degree of interaction, sampling the input along with characterizing the uncertainty of corresponding output. Moreover, authoring and screen casting tools play a vital role too.

Acknowledgement. Ms. Eva Kaushik*, scholar, Dr. Akhilesh Das Gupta Institute of Technology & Management remarkably enhanced our manuscript. Her valuable contribution and endeavor is commendable.

References

1. Pan, H.: Morphological simulation of phyllotactic spiral-ring patterns by the galactic spiral equations from the ROTASE model (2022). https://doi.org/10.20944/preprints202208.0231.v1
2. Schwabe, B., Niemeyer, J.C.: Deep zoom-in simulation of a fuzzy dark matter galactic halo. Phys. Rev. Lett. **128**(18) (2022). https://doi.org/10.1103/physrevlett.128.181301
3. Morphological simulation of phyllotactic spiral-ring patterns by the galactic spiral equations from the ROTASE model. Adv. Image Video Process. **10**(4) (2022). https://doi.org/10.14738/aivp.104.12835
4. Hodgson, T., Vazza, F., Johnston-Hollitt, M., McKinley, B.: Figaro Simulation: Filaments & Galactic Radio Simulation, vol. 38. Publications of the Astronomical Society of Australia (2021). https://doi.org/10.1017/pasa.2021.32
5. The Galactic Center: The Galactic Supermassive Black Hole, pp. 1–24 (2020). https://doi.org/10.2307/j.ctv17db3hw.5
6. Schneider, E.E., Robertson, B.E.: Introducing CGOLS: the cholla galactic outflow simulation suite. Astrophys. J.. J. **860**(2), 135 (2018). https://doi.org/10.3847/1538-4357/aac329
7. Michikoshi, S., Kokubo, E.: Global N-body simulation of galactic spiral arms. Mon. Not. R. Astron. Soc. **481**(1), 185–193 (2018). https://doi.org/10.1093/mnras/sty2274

8. Aoyama, S., et al.: Galaxy simulation with dust formation and destruction. Mon. Not. R. Astron. Soc. **466**(1), 105–121 (2017)

9. Keller, B.W., Wadsley, J., Benincasa, S.M., Couchman, H.M.P.: A superbubble feedback model for galaxy simulations. Mon. Not. R. Astron. Soc. **442**(4), 3013–3025 (2014)

10. Baxter, E.J., Dodelson, S., Koushiappas, S.M., Strigari, L.E.: Constraining dark matter in galactic substructure. Phys. Rev. D **82**(12), 123511 (2010)

11. Majewski, S.R., et al.: Galactic dynamics and local dark matter. arXiv preprint arXiv:0902. 2759 (2009)

12. Shirokova, K.S.: Simulation of the Motion of Stars Escaping from the Galactic Center." Open Astronomy **24**(4) (2015). https://doi.org/10.1515/astro-2017-0246

13. Bannikova, E.Y., Vakulik, V.G., Sergeev, A.V.: N-body simulation of a clumpy torus: application to active galactic nuclei. Mon. Not. R. Astron. Soc. **424**(2), 820–829 (2012). https://doi.org/10.1111/j.1365-2966.2012.21186.x

Protocol Anomaly Detection in IIoT

S. S. Prasanna[✉], G. S. R. Emil Selvan, and M. P. Ramkumar

Department of Computer Science and Engineering, Thiagarajar College of Engineering,
Madurai, India
ssprasannamdu@gmail.com, {emil,ramkumar}@tce.edu

Abstract. Industrial IoT (IIoT) belongs to the category of Operational Technology (OT) network, which is different from Information Technology (IT) network. The latter can be infused with large amount of computing power whereas the former lacks the support of holding that much computing power since OT network is a resource-constrained area. As a result, a lightweight framework must be built to improve and extend the behavior and performance of the OT network. However, if resource usage is low, the performance will be degraded. A combination of LightGBM and Stochastic Gradient Descent algorithms for anomaly detection is proposed. In the first part, the feature selection process using LightGBM algorithm is discussed and then the model training and anomaly detection methodology using Stochastic Gradient Descent are illustrated. Also, the drawbacks of Gradient Descent algorithm and how the SGD overcomes those drawbacks are discussed. The findings from experiments show that the suggested model generated high throughput and high anomaly detection accuracy.

Keywords: IIoT · Protocol Anomaly Detection · Stochastic Gradient Descent · LightGBM · Anomaly Detection

1 Introduction

Anomaly refers to a situation where something is partially or completely different from the normal situation. Technically, anomaly detection refers to the process of detecting abnormal behaviors or strange performance of the network. It is different from signature-based detection method. The latter needs a database of previously identified attack signatures. Rather, the former needs a training to detect the anomaly behaviors of network

Protocol Anomaly Detection, also known as Network Behavior Anomaly Detection (NBAD) is a variant of anomaly detection. Every protocol will have a standard behavior, known as RFC standard (Request for Comment). In this method of detection, the packets that violates the RFC standard are flagged as threat. Three major network behavior components are monitored – traffic flow patterns, passive traffic analysis and network performance. The primary goal of protocol anomaly detection is to identify the packets that are different from the normal network traffic behavior. Nowadays, this protocol anomaly detection becomes an important or base component of intrusion detection systems and intrusion prevention systems. This protocol anomaly detection method is best suited for IIoT network as it lacks support for performing complex operations and storing huge data. It is enough to identify whether the network traffic flow is within the behavioral limits. Because the attack traffic will not fit in the RFC standard. The attack itself is the process of violating the protocol standards and establishing an illegal connection with the target.

One of the main objectives for any algorithm is to reduce the error and generalize, when a new data is appended. This can be achieved through implementing optimization techniques. Gradient Descent is an iterative un-constrained optimization technique. Large dataset is needed for good generalization which further increases the computational cost, i.e. as the dataset size increases, the computational cost and time will be huge for computing single gradient step. Gradient Descent is often considered as unreliable and slow. Stochastic Gradient Descent (SGD) works well and attains the local minimum in acceptable time. In addition to that, Gradient Descent can't handle the non-convex problems, where SGD overcomes that issue.

2 Literature Survey

Security incidents are decreased by a novel hinge classification technique based on mini-batch gradient descent with an adaptive learning rate and momentum (HCA-MBGDALRM) that is proposed in the work [2]. Since IIoT environment produces huge amount of data, it is impractical to verify that each packet contains legitimate data or attack data. Hence, a parallel framework for HCA-MBGDALRM is also implemented along with it, which accelerates the processing speed of large traffic. In [4] SYN flood attack is detected using anomaly detection. Since SYN flood attack is related to TCP protocol, the TCP header is checked against the RFC standard. TCP header includes IP header length, Type of service, Flags, etc. A proposal to combine Chi-square distance into markov chain method is made to detect protocol anomalies [5]. Another method [6] uses finite state machine and extended finite state machine for detecting anomalies in Neighbor Discovery Protocol (NDP) based on strict anomaly detection. The work presented in [7] describes the need of a lightweight and fast accessing Network Intrusion Detection System (NIDS). The proposed NIDS uses Support Vector Machine (SVM) for anomalies classification.

A reliable and resource-efficient anomaly detection method for IIoT is presented [9]. Two stages of feature extraction make up this system for unsupervised anomaly detection – one with autoencoder and another with efficient deepexplainer. Then a model is developed based on efficient deepexplainer which detects the anomaly. A new LGBM-infused feature selection model named EFS-LGBM is proposed [9]. The work presented in [11] uses LGBM as feature selection algorithm. The work done in [12] uses SGD as one of the proposed ML algorithm for spam detection. Also, it states that SGD is capable of handling large-scale datasets. The purpose of anomaly point detection is to identify outliers in dataset within in given time. So, a hierarchical representation approach for anomaly detection is introduced [13]. For deep anomaly detection in the IIoT, a communication-efficient Federated Learning framework is presented [14]. The Attention Mechanism based Convolutional Neural Network-Long Short-Term Memory

(AMCNN-LSTM) is used in this unsupervised framework. The works depicted in [15] uses graph neural networks for anomaly detection. This study also examined three case studies related to smart transportation, smart energy and smart factory. An approach for identifying anomalous behavior was developed [17], which makes use of Hotelling's T statistics as well as multivariate statistical analysis.

According to the research in [18], anomaly detection is the method of identifying patterns in the network that deviate from the expected flow. The primary objective of anomaly detection is to identify any network flow that is abnormal and label it as such. This paper also discusses the difficulties in performing such activity. It will be challenging to define normal regions, because normal behavior may change over time and varies in different domains. It is suggested to use a hybrid deep-learning system for anomaly detection [19]. The anomaly detection module, which is the initial step, employs the restricted Boltzmann machine and gradient-based support vector machines to detect unusual activity. The most important aspects that contributes to anomaly detection are summarized in the work [20]. A detailed background study and anomaly detection techniques are presented. It states that anomaly-based detection has ability to identify both known and undiscovered attacks and anomalies as the detection is performed based on unusual patterns. Another work proposes 2 approaches for anomaly detection [23]. First is based on graph influenced by finite state automaton and the second is a proprietary algorithm based on cyclic lists. A new intelligent platform named Hybrid Anomaly Detection Model (HADM) is proposed and the performance evaluation of several combinations of algorithms with HADM against different datasets are compared [21].

TCP SYN flood attack is a kind of DoS attack which targets TCP three-way handshake process. An anomaly-based detection algorithm was developed [3] to detect this attack. The alert is triggered if the TCP header size is less than or equal to 20 bytes or if the packet contains unwanted TCP flags. The anomaly detection approach proposed in [8] employs 3 clustering methods - fuzzy c-means, density-based and k-means. A new method named DeepStack-DTI is introduced in [10] where LightGBM is used for feature selection and a five-layered validation units including SVM, XGBoost, Logistic Regression, DNN and GRU. In the context of IIoT aging, a methodology has been developed [16] for abnormal behavior detection using PCA, Hotelling's T statistics and univariate cumulative sum. The work in [22] proposes a binary and a multi-class anomaly classification models. The binary model uses LSTM, Bi-LSTM and GRU techniques and the multi-class model uses CNN and RNN for classification.

3 LightGBM

Feature selection is essential in machine learning. Since there may be some irrelevant attributes present in the dataset, it is necessary to remove those. By doing so, training and prediction time can be reduced and can make the model less prone to noise-based predictions.

Light Gradient Boosting Machine (or) LightGBM (or) LGBM is one of the gradient boosting algorithms which learns through tree-based method. It is similar to XGBoost but varies in some aspects. Instead of growing horizontally, LGBM grows vertically, i.e., it grows tree leaf-wise whereas others grow level-wise. The difference in tree growth of

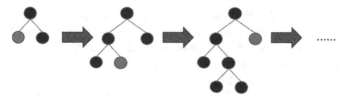

Fig. 1. Leaf-wise growth (LightGBM)

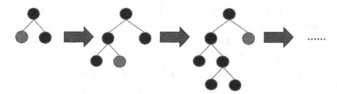

Fig. 2. Level-wise growth (XGBoost)

LGBM and XGBoost are picturized in Figs 1 and 2 respectively. Due to this feature, the loss in sequential boosting process is reduced. LGBM yields higher accuracy than other boosting algorithms. The feature_importances_attribute function contained in LGBM, as in XGBoost, is used to determine the importance of each feature.

4   Stochastic Gradient Descent

Generally gradient descent is a generic optimization algorithm, proficient in finding optimal solutions to broad- ranging problems. By tweaking parameters iteratively, the cost can be minimized. Among the 3 types of gradient descent, Stochastic Gradient Descent (SGD) has been chosen for training this model. As per the name "Stochastic", the SGD calculates the derivative of one random data point's loss instead of calculating it for all data points. The term "Stochastic" means random. This feature makes SGD to perform faster than traditional gradient descent. Also, SGB is good in handling huge amount of input data, where traditional gradient descent is slow on that. Scikit-learn provides SGDClassifier module for solving classification problems. One of the parameters of SGDClassifier is alpha. This alpha is the tuning parameter that determines how much the model may be penalized, which is a constant that multiplies the regularization term. By default, alpha value is set to 0.0001. The randomness can be induced in selecting data points. At each iteration, SGD randomly picks one data point from the whole dataset and thereby reducing the computations.

A single real integer is used to represent the cost function, which measures the difference between actual and anticipated values at the current position. This cost function is produced when a hypothesis is formed with initial parameters, and these parameters are modified through known data using gradient descent techniques to minimize the cost function. SGD is useful for both locating the global minimum and evading the local minimum. SGD is simpler to allocate to a specific location in memory. It is more efficient in handling large amount of datasets and it is relatively faster than other gradient descent algorithms in terms of computation.

5 Proposed Methodology

The proposed framework divided into 3 stages – Feature selection, Training and Testing/Evaluation phase. The proposed framework gets the input as CSV file which contains the necessary packet header attributes. Then, irrelevant features are identified and eliminated in feature selection stage. With the relevant features, training and testing stages are executed. The flow diagram is represented in Fig 3.

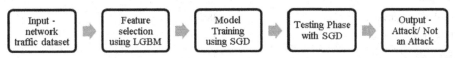

Fig. 3. Proposed process flow

5.1 Dataset

X-IIoTID dataset [1] from IEEE DataPort is used. It consists of 65 features out of which 5 features (Date, Timestamp, SrcIP, DstIP, Service) are excluded. It contains output class into 3 types. As a preprocessing work, those 3 columns are converged into single column comprising of either Normal or Attack as output. The further processes are executed with the remaining 60 features and 1 target column.

5.2 Feature Selection

Selecting the feature that are closely related to the output value. For this, LGBM is used. LGBM output is ordered in descending order of their importance value. The one with lowest importance value is removed from the dataset. Because, by including that feature, the model will get trained in a wrong manner or maybe get confused and make decisions.

5.3 Model Training

Model training is carried out with Stochastic Gradient Descent. It is generally used for function approximation and classification problems. Hence, we are using this neural network for classification problem. The final stage is testing. Testing is also done using SGD. While comparing SGD with other neural networks for classification problem, SGD achieves greater results than others.

6 Results and Discussion

The dataset has to be preprocessed before beginning feature selection phase. Empty cells and cells with value hyphen (–) are filled with zero. Boolean values are replaced with zeros and ones. Protocols – UDP, TCP and ICMP are replaced with the numbers 0, 1 and 2 respectively. By doing these tasks, the dataset will now contain only numeric values which indicates the ready-to-use state of a dataset. The feature selection results in 5 features that are irrelevant to the target column. The feature importance values for each feature are tabulated in Table 1. Before initiating the model training, those 5 features are eliminated from the dataset.

Table 1. Feature Selection

Feature	Feature Importance Value
Des_port	357
Avg_num_cswch/s	267
Duration	262
Scr_pkts	142
read_write_physical.process	138
Des_bytes	136
Scr_port	134
Scr_bytes	126
Avg_num_Proc/s	106
Des_ip_bytes	93
anomaly_alert	87
Paket_rate	79
Scr_ip_bytes	69
is_syn_only	65
Scr_bytes_ratio	65
Std_num_proc/s	56
Des_bytes_ratio	56
Login_attempt	51
Avg_nice_time	51
Des_pkts_ratio	50
byte_rate	50
Conn_state	49
Total_bytes	39
FIN or RST	38
is_with_payload	38
OSSEC_alert_level	28
Avg_user_time	27
Scr_packts_ratio	25
Avg_wtps	25
Protocol	23
Avg_system_time	22
Std_num_cswch/s	20

(*continued*)

Table 1. (*continued*)

Feature	Feature Importance Value
is_SYN_ACK	18
total_packet	17
Avg_ldavg_1	17
Avg_kbmemused	16
Std_ldavg_1	15
Des_pkts	15
Avg_rtps	14
Succesful_login	12
Std_user_time	12
Avg_iowait_time	10
OSSEC_alert	9
Process_activity	9
Std_kbmemused	8
Avg_tps	8
Std_nice_time	8
Std_tps	7
Std_system_time	7
is_pure_ack	7
File_activity	4
Std_rtps	4
Std_wtps	3
Std_ideal_time	3
Avg_ideal_time	3

Before the training phase, the dataset is divided into two halves such that the ratio of train-to-test data is 7:3. For evaluating the training process of the model, each time it has to be trained from scratch, without utilizing the training results from previous attempts. This process is termed as cross validation. The conclusion of better model can be driven from the result of this cross-validation process. The learning curve, which depicts how well the model is trained, is calculated using the training dataset. The cross validation learning curve, which is created from the validation dataset and shows how effectively the model generalizes, Fig 4 shows the learning curve for SGDClassifier. This learning curve is developed with yellowbrick package. From the graph, we can infer that the cross-validation score is below the training score and both the scores are higher. This indicates that the model doesn't result in overfitting or underfitting.

After training the SGD model with alpha value 0.01, a completely new set of data is given for testing the model. To evaluate the effectiveness of the model, a confusion matrix

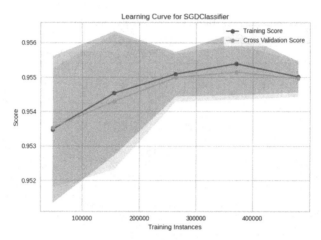

Fig. 4. Learning Curve

is utilized. This matrix is divided into 2 dimensions as the output value is either attack or normal. This matrix shows how good the model does the prediction or classification work.

- True Positive (TP): Predicting true as true
- True Negative (TN): Predicting false as false
- False Positive (FP): Predicting false as true
- False Negative (FN): Predicting true as false

Figure 5 illustrates the confusion matrix. The confusion matrix indicates that the TP (126,224) and TN (43,085) counts are much higher than the FN (217) and FP (76,480) counts, which implies that the model performs better. Accuracy is the calculation of true class predictions over total data, i.e. (TP+TN)/total observations. The proposed model outperforms other models discussed in Sect. 2 and yields the accuracy score of 93.50%.

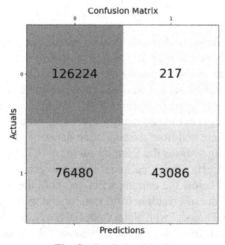

Fig. 5. Confusion Matrix

7 Conclusion

Since IIoT is a resource-constrained environment, training and classification algorithms should produce higher accuracy by using fewer resources. We must implement an algorithm that meets the requirements of minimal resource consumption and high accuracy because resource usage directly influences algorithm performance. SGD satisfies both the conditions. As SGD uses single training sample, it is easier to fit in the memory and is computationally fast in handling large datasets. A protocol anomaly detection framework is proposed which uses a combination of LGBM and Stochastic Gradient Descent algorithms. The experimental results demonstrate that the proposed model beats alternative approaches by achieving accuracy of 93.50%.

References

1. Al-Hawawreh, M., Sitnikova, E., Aboutorab, N.: X-IIoTID: A Connectivity- and Device-agnostic Intrusion Dataset for Industrial Internet of Things. IEEE Dataport (2021). https://doi.org/10.21227/mpb6-py55
2. Yan, X., et al.: Trustworthy network anomaly detection based on an adaptive learning rate and momentum in IIoT. IEEE Trans. Indust. Inform. **16.9**, 6182–6192 (2020)
3. Haris, S.H.C., Ahmad, R.B., Ghani, M.A.H.A.: Detecting TCP SYN flood attack based on anomaly detection. In: 2010 Second International Conference on Network Applications, Protocols and Services. IEEE (2010)
4. Qin, Z., et al.: Improvement of protocol anomaly detection based on markov chain and its application. In: Chen, G., Pan, Y., Guo, M., Lu, J. (eds.) Parallel and Distributed Processing and Applications - ISPA 2005 Workshops. ISPA 2005. LNCS, vol. 3759. Springer, Berlin, Heidelberg (2005). https://doi.org/10.1007/11576259_43
5. Firas, N., Kadhum, M., El-Taj, H.: Neighbor discovery protocol anomaly detection using finite state machine and strict anomaly detection. In: Proceedings of the 4th International Conference on Internet Applications, Protocols and Services (NETAPPS2015) (2015)
6. Priya, V., et al.: Robust attack detection approach for IIoT using ensemble classifier. arXiv preprint arXiv:2102.01515 (2021)
7. Huang, Z., et al.: An energy-efficient and trustworthy unsupervised anomaly detection framework (EATU) for IIoT. ACM Trans. Sensor Networks **18.4**, 1–18 (2022)
8. Hore, Umesh W., and D. G. Wakde. "An Effective Approach of IIoT for Anomaly Detection Using Unsupervised Machine Learning Approach." J. IoT Soc. Mob. Anal. Cloud 4 (2022): 184–197
9. Zhou, K., et al.: Fast prediction of reservoir permeability based on embedded feature selection and LightGBM using direct logging data. Measure. Sci. Technol. **31.4**, 045101 (2020)
10. Zhang, Y., et al.: DeepStack-DTIs: predicting drug–target interactions using LightGBM feature selection and deep-stacked ensemble classifier. Interdiscipl. Sci. Comput. Life Sci. 1–20 (2022)
11. Elshoush, Huwaida T., Dinar, E.A.: Using adaboost and stochastic gradient descent (sgd) algorithms with R and orange software for filtering e-mail spam. In: 2019 11th comPuter Science and Electronic Engineering (CEEC). IEEE (2019)
12. Zhan, P., et al.: Temporal anomaly detection on IIoT-enabled manufacturing. J. Intell. Manufac. **32**, 1669–1678 (2021)
13. Liu, Yi, et al. "Deep anomaly detection for time-series data in industrial IoT: A communication-efficient on-device federated learning approach." IEEE Internet of Things Journal 8.8 (2020): 6348–6358

14. Wu, Y., Dai, H.-N., Tang, H.: Graph neural networks for anomaly detection in industrial internet of things. IEEE Internet Things J. **9**(12), 9214–9231 (2021)
15. Aoudi, W., Almgren, M.: A scalable specification-agnostic multi-sensor anomaly detection system for IIoT environments. Int. J. Crit. Infrastruct. Prot. **30**, 100377 (2020)
16. Genge, B., Haller, P., Enăchescu, C.: Anomaly detection in aging industrial internet of things. IEEE Access **7**, 74217–74230 (2019)
17. Sharghivand, N., Derakhshan, F.: Classification and intelligent mining of anomalies in Industrial IoT. In: Karimipour, H., Derakhshan, F. (eds.) AI-Enabled Threat Detection and Security Analysis for Industrial IoT, pp. 163–180. Springer International Publishing, Cham (2021). https://doi.org/10.1007/978-3-030-76613-9_9
18. Garg, S., et al.: Hybrid deep-learning-based anomaly detection scheme for suspicious flow detection in SDN: a social multimedia perspective. IEEE Trans. Multimedia **21.3**, 566–578 (2019)
19. Gilberto, F., et al.: A comprehensive survey on network anomaly detection. Telecommun. Syst. **70**, 447–489 (2019)
20. Milosz, S., et al.: Anomaly detection in cyclic communication in OT protocols. Energies **15.4**, 1517 (2022)
21. Mehrnoosh, M., et al.: Performance evaluation of a combined anomaly detection platform. IEEE Access **7**, 100964–100978 (2019)
22. Ullah, I., Mahmoud, Q.H.: Design and development of RNN anomaly detection model for IoT networks. IEEE Access **10**, 62722–62750 (2022)
23. Jiang, J.-R., Chen, Y.-T.: Industrial control system anomaly detection and classification based on network traffic. IEEE Access **10**, 41874–41888 (2022)

Agricultural Informatics & ICT: The Foundation, Issues, Challenges and Possible Solutions—*A Policy Work*

P. K. Paul[1] ⬤, Mustafa Kayyali[2(✉)] ⬤, and Ricardo Saavedra[2] ⬤

[1] Department of CIS, Raiganj University, Raiganj, India
[2] Azteca University, Chalco de Díaz Covarrubias, Mexico
`kayyalimustafa@gmail.com`, `international@univ-azteca.edu.mx`

Abstract. Informatics is an interdisciplinary and valuable field of study dedicated in various information related tasks viz. collecting, selecting, organizing, processing, managing as well as disseminating of contents and information using appropriate technologies. The application and merging of Informatics with other domains and fields have been development some of the other domains and fields like Bio Informatics, Health Informatics, Geo Informatics, Agricultural Informatics, and so on. Further, Agricultural Informatics or Agro ICT is the merger of 'Agricultural Science' and 'Informatics/Information Science'. Agricultural Informatics is dedicated to designing, developing and modernizing Agricultural systems and practices with the help of ICT. Various technologies have already being used in agricultural activities such as basic computing tools, and also other IT components viz. Software Technologies, Network Technologies, Multimedia Technologies, Web Technologies, Database Technologies in recent past, and many other emerging technologies also employing in better and healthy Agricultural Practice. Furthermore, in the current context, many countries have initiated ICT in Agricultural Systems and there are progressive scenarios. Though some of the issues and challenges are also being treated as an obstacle in certain contexts and this work is about the Agricultural Informatics with reference to the importance and finding the issues, challenges of ICT in Agricultural Systems with proper solutions in Agricultural Systems.

Keywords: Agricultural Informatics · Agro ICT · Digital Agriculture · Smart Agriculture · Development Studies

1 Introduction

The main feature of Agricultural Informatics is it is not only modern field of study but also very much dynamic, and being associated with the technologies, information and various concern of agriculture. As a field of study, it is called as Agricultural Informatics but in general, it is also referred to as ICT Applications in Agriculture, IT in Agricultural Development. Some of the common nomenclature and name of Agricultural Informatics are includes AIT (Agricultural Information Technology), Agricultural Information

P. Das et al. (Eds.): AMRIT 2023, CCIS 1954, pp. 47–60, 2024.
https://doi.org/10.1007/978-3-031-47221-3_5

Systems (AIS), and Agricultural Communication Technology (ACT), etc. Agricultural Informatics is talked about the applications of the Informatics and IT applications in the Agriculture and allied fields [4, 29]. Agricultural Informatics, therefore, uses various technologies of Informatics and IT components viz.—

- Software Technologies
- Network Technologies
- Database Technologies
- Web Technologies
- Multimedia Technologies and so on.

Internationally not only developed countries but also many developing countries have started practice with the Agricultural Informatics. Many organizations, ministry of many Governments, institutions, agro related associations and firms are engaged with the Agricultural Informatics practice or ICT Applications in Agriculture and allied domains. Though it has many benefits but it also concerning with various issues and challenges [6, 11, 38]. Overcoming of such issues can lead to a developed and sustainable agro system perfectly. Agricultural Informatics is an amazing interdisciplinary professional subject dedicated in development of proper agricultural systems for the modern and smarter agricultural practice.

2   Objective and Novelty of the Work

Most of the existing work is done related to the Agricultural Informatics or Agricultural Information Technology related to design, developing a model on smart agricultural production, smart agricultural development, and smart crop production while present work entitled 'Agricultural Informatics & ICT: The foundation, issues, challenges and possible solutions—*A Policy work*' is a policy work and completely theoretical in nature. Further this work is mainly dedicated to the following (but not limited to)—

- To gather about the foundation of Agricultural Informatics emphasizing its core features and attributes briefly.
- To find-out the importance and significance of developing digital or smart agricultural systems using Agro Informatics.
- To find out the issues and core challenges in respect of practicing Agricultural Informatics including other subjects dedicated in ICT/ IT in the context of Agricultural Sector.
- To learn the possible suggestions in regard to the core issues and challenges better and effective practice and uses of Agricultural Informatics in the light of developed and developing countries.

Therefore this work is suitable for the educationalist, administrators, policy makers and researchers to gather information about the Agricultural Informatics specially on issues and possible solutions.

2.1 Methodology

The proposed work is based existing work and therefore theoretical in nature. This work is conceptual and prepared upon studying existing topics and research work and thus different secondary sources have been analyzed and reported here. Various journals related to the agricultural sciences, Agricultural Informatics, ICT and Computing Applications have been analyzed and reported. Review of last five year's work specially put on importance while preparing this research work. General Google search also being used while doing this work, and for this common keywords considered as 'Agricultural Informatics', 'ICT in Agriculture', 'Issues in Agro Informatics'.

3   Agricultural Informatics: The Foundation and Enhancement

Agriculture is valuable and important everywhere. Agriculture also has an economic role and treated as a tool for development and in this regard, Agricultural Informatics is dedicated to healthy ICT based Agriculture Sector promotion. Initially, Agricultural Informatics meant for only Computer and ICT Applications in agricultural information management but now it is beyond such practices and include applications of various sub-areas of emerging technologies (of ICT) viz.—

- Data Analytics and Big Data Management.
- Virtualization and Cloud Systems.
- Effective Statistical Systems and Applications.
- Internet of Things, and Web of Things.
- Advanced and Intelligent Converged Network.
- Usability Engineering and UXD.
- Interaction Techniques with 3D Graphics.
- HCI and Human Centered Computing.
- Wireless Sensor Network (WSN)
- Multimedia Systems and Database
- Effective and advanced Satellite communication and Technology etc.

Therefore this is in many universities, Agro Informatics and allied branches become a program of study, internationally with Bachelors, Masters, and Doctoral Degrees. The most common nomenclatures are Agricultural Informatics, Agricultural Information Technology [1, 9, 33].

Herewith the help of Agricultural Informatics developing and undeveloped countries may engage in employment generation. Farmers are suffered from various aspects in respect of Agriculture such as heating, flood, cold, situation of drought, managing and assuring from the insect, proper pest infestations, and managing and finding diseases from the general weather and also from the climate change, etc. and in this regard, Agricultural Informatics practice can be considered as important and valuable. Skilled manpower in modern agriculture or cultivation methods is the need of the hour with reference to the skills of Computing and IT. Agro Informatics is nicely effective in modern agricultural practice with modern computational support and indirectly can be helpful in various aspects viz. in security of the food, proper and actual need of the nutrition, in managing effective global trade, in advanced technological development and systems, ecological and environmental concern; and Agro Informatics is needed in directly and indirectly. Strategically Agro Informatics is needed in strengthening agricultural systems and its development—

- Input of Agricultural Systems
- Output of the Agricultural practices and systems
- Proper and effective Agricultural System's integration
- For effective Agricultural Business and Marketing
- For Healthy Agricultural Systems development including Post Production
- In Easiest Agro Transportation
- For the concern of real and required concern of security of the foods.
- For effective and advancing in Agricultural Systems development
- In better Agro based Value-chain development.
- Regarding proper and healthy climate related concern [2, 10, 31].

Various technological subjects and fields like ICT, Informatics, Information Science and Technology, IT, Advanced Computation Sciences, etc. are important in effective Agricultural Informatics practices and development. With the Agro Informatics agriculture systems development can be developed effectively with pre production and post production activities. Further, it is needed in the healthy development of the global innovations in the field of agriculture including its productivity, effective economic development including social concern, etc. IT/ Computing is a core of the Agricultural Informatics and therefore educated in the field will be able in various other jobs related to the science or technology. Therefore it is the high time for the healthy agricultural dynamism in the country. Agricultural Informatics including other allied branches, Agriculture Sciences with knowledge and skilling in Agriculture is the need of the hour and required for the healthy development of Agricultural Systems. Below mentioned areas are important in effective healthy Agricultural Informatics promotion—

- Proper skilling and knowledge of emerging technologies.
- Agriculture related knowledge.
- Communication for effective pre and post Agro activities.
- Entrepreneurship skills for Agro Business.
- Required leadership tech and managerial aspects of Agriculture Management and allied areas, etc.

Agricultural Informatics is dedicated to Increases *Efficiency* using proper and effective monitoring of farming and cultivation including agricultural production management due to real time data and information gathering [7, 20, 36].

Expansion is another benefit of the Agricultural Systems and also in smart closed-cycle agricultural systems; and in Cleaning and Purity of Agro Space, Agro Informatics is a good tool for the development of the agricultural activities such as managing pesticides and fertilizers, etc. Agricultural Informatics required in *Quicker Agricultural Systems* using robotics, artificial intelligence, data analytics, etc. As far as *Healthy & Quality Production of Agriculture* is concerned Agro Informatics is the need of the hour due to uses of high end IT and Computing support viz. Aerial drone monitoring systems, etc. The *Livestock Management* is helpful in some of the activities of agro post production activities such as marketing, agro business promotion, etc. [3, 13, 37].

4   Core Issues

As an interdisciplinary field of study and practice Agricultural Informatics, though it offers many opportunities but at the same time it is concern with huge challenges and issues. Though, with solving such issues Agricultural sector can be more developed, intelligent and prosperous. The following are major issues (also refer Fig. 2) mentioned.

4.1   Education and Manpower: Issues

Agricultural Informatics is field interdisciplinary in nature and practicing domain and composed of various areas viz. Agricultural Science, Information Science, Computing and Management. Further Agricultural Informatics also deals with the aspects of the Environment, Sociology and Mathematical Sciences. But it is worthy to note that the proper manpower in the field is still limited than the requirement [12, 18, 39]. Only a few universities offer degrees and the majority are in Science and Engineering streams. Additionally, there is a limitation in proper research manpower in Agricultural Informatics and it is throughout the world. Moreover, there is a limitation in skilled manpower in the field with basic operational activities of Agricultural Information Technology.

4.2   Implementation of the Technology: Issues

Agricultural Informatics is technologically focused and here apart from basic technologies other emerging technologies also considered as important; and such technologies changing rapidly and therefore there is an issue of proper implementation of such technologies into existing systems. Among the technologies, a few important are Big Data Systems, Virtualization and Cloud Systems, Statistical Systems and Management, Web of Things and Internet of Things, Advanced Converged Network, Wireless Sensor Network, UXD and Usability Engineering, Usability Systems and Interaction Systems, HCI and Human Centered Computing, Multimedia Systems and Database and so on. Proper integration is an important issue to look after [5, 14, 35].

4.3   Financial and Monetary: Issues

Finance is the most important and valuable in all kinds of contexts and sectors. Regarding Agricultural sector and space is concerned proper and required financial involvement is important in different stages including crops, harvesting, plants, irrigation, soil and land management, marketing and agro business and so on. Apart from these in manpower billing also finance is most important. Agricultural Informatics and Agro ICT is concerned with initial financial involvement in technology and equipment purchasing, technological maintenance, skill development and training program initiation, technology transfer and so on [5, 40, 41]. The emerging technologies need costly software and hardware involvement and therefore proper budgeting is required. But in developing countries, small firms and low and financially weak farmers—this is an important issue.

4.4 Digital Divide: Issues

Digital Divide is about the divide of the technology in respect of tools, services, equipment, technologies among the have and haven't. Digital Divide is an important issue in Agricultural Informatics as it involves with huge digital and technological involvement. In many countries, it is an important issue that due to the already alarming Digital Divide issue the Agricultural Informatics program been unsuccessful [8, 15].

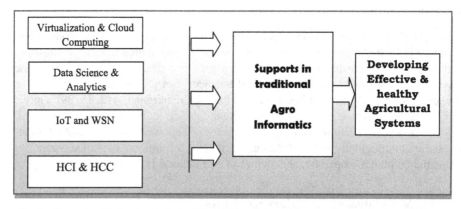

Fig. 1. Emerging Technologies in traditional Agro IT resulting more advance Agro Systems

In developing countries, it is an important issue. The less availability of the technologies, products, IT services viz. Internet connections, broadband, etc. also Agricultural Informatics initiation becomes difficult [17, 32]. Refer Fig. 1 regarding their role into Agriculture [34].

4.5 Government Support: Issues

The Governmental Support is very important and urgent for fulfillment of any mission, agenda, scheme and technological implementation in anykind of sector, and regarding Agricultural Informatics support from the Government is important and it is noted that in technological implementation, in proper funding and budget allocation, in technology transfer, in proper policy formulation and their solid implementation there are lack of Governmental support.

4.6 Awareness and Literacy on Agro Informatics: The Issues

As Agricultural Informatics not only a practicing field but also an interdisciplinary subject and therefore proper awareness is the need of hour and required. Agricultural Informatics holds various benefits and advantages and key mover in smarter agricultural development and digital agricultural development. The lack of awareness is an important issue and here it is worthy to note that proper and lack of initiative in this regard. The lack of initiative is noted from the Government, Farming Companies, Farmers, Agricultural Associations, etc. [16, 22, 28].

4.7   Skill Development: Issues

Agricultural Informatics is changing rapidly. Initially only core technologies are considered important in healthy Agricultural Informatics practice but gradually other technologies have emerged such as Data Analytics and Big Data Management, Virtualization and Cloud Systems., Effective Statistical Systems and Applications, Internet of Things, and Web of Things, Advanced and Intelligent Converged Network, Usability Engineering and UXD, Interaction Techniques with 3D Graphics, HCI and Human Centered Computing, Wireless Sensor Network (WSN), Multimedia Systems and Database, Effective and advanced Satellite communication and Technology. Therefore, there is a lack of Agricultural Informatics practice with such technologies. Further the skill development program for the farmers, re-enhancement and skilling also important concern and there is a gap on such in developing countries and in some context in the developed countries as well. The short term program, training program are limited in most of countries according to the study [19, 21, 34].

Fig. 2. Major Issues in healthy Agro ICT practice

4.8   Environment, Locality and Awareness: Issues

Agricultural Informatics is associated with various issues related with the environment and ecology. Since Agriculture is connected with the environment therefore Agricultural Informatics deals with various issues on environment viz. wider uses of computing devices, harmfulness of the IT and Computing devices. There are issues like localization, awareness in community and root level of the society also important issue noted according to this study.

4.9   Policies, Framework and Regulations: Issues

Various technologies and systems of Agricultural Informatics required in Agriculture, Environment, Society and so on. Though the shortage of policies, rules and regulations are can be noted with important concern. The issues can be noted with Governmental Ministries, Departments, Educational Institutes and Universities, Research Centers, Agricultural Associations in respect of proper policies, guidelines, and frameworks.

4.10   Existing and New Technological Integration: Issues

Some of the technologies such as Software Systems & Technologies, Network Systems & Technologies, Multimedia Systems & Technologies, Web Systems & Technologies, Database Systems & Technologies treated as most vital. Though the issues are also be noted on emerging technologies on finance, technological integration, proper technical knowledge on uses and implementation, and so on.

4.11   Technological Development and Cycling: Issues

Emerging technologies such as Data Analytics and Big Data Management, Virtualization and Cloud Systems, Internet of Things, and Web of Things, Advanced Converged Network, Usability Engineering and UXD, Interaction Techniques with 3D Graphics, HCI and Human Centered Computing, etc. needs proper cycling and uses. But such issues not only noted in the developing world but also in the developed world in some context [7, 23, 27].

5   Proposed Solutions

Agricultural Informatics since a broad, interdisciplinary field and practicing area and rising internationally, therefore, increasing its uses. It is a fact that the issues can be solved with proper remedies and therefore herewith few suggesting measure and possibilities are mentioned as under (also refer Fig: 3)—

5.1 Education and Manpower: Proposal

Agricultural Informatics program can be started easily with proper initiation at different existing levels at the different universities where computing, IT and other allied branches exit as a specialization. Further, Agricultural Informatics can also start in agricultural sciences, ecology and environment, horticulture related departments as a specialization of Agricultural Informatics such as AIT (i.e. Agricultural Information Technology), Agricultural Information Systems, Smart and Digital Agriculture and so on. Further in other streams viz. Management, Commerce and Business also Agricultural Informatics can be started with a special focus on Agricultural Information Technology Management. For the low end personnel and farmers short term skill based certificates, diplomas, etc. can be started.

5.2 Implementation of the Technology: Proposal

The involvement of the technologies in Agricultural Informatics practice can be solved out by using proper strategies and feedback. Different kinds of technological implementation can be solved with the initiation of the proper patches, firewalls, utilities, proper evaluation also consider as important in healthy Agricultural Informatics practice. Emerging technologies viz. Data Analytics and Big Data Management, Virtualization and Cloud Systems, Internet of Things, and Web of Things, Advanced Converged Network, Usability Engineering and UXD etc. need to implement accordingly and as per changing trends with proper and skilled manpower only.

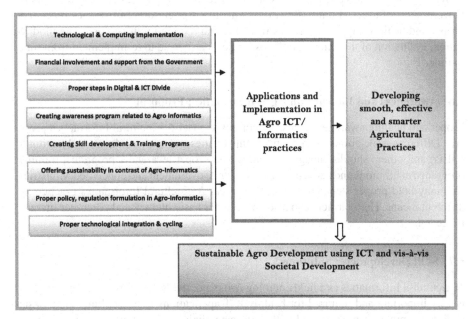

Fig. 3. Proper solutions to the issues of Agro Informatics practices in Developing Countries

5.3 Financial and Monetary: Proposal

The issues related to finance can be solved with proper financial support from the Government, Associations and Organizations related to Agriculture, Development and Technologies. Furthermore, budget allocation is also important and urgent in solving financial issues. Developing countries need proper steps and policy formulation regarding technological purchasing and others involved with the Agricultural Informatics. The loan facilities can also be good solutions from the concerned or allied ministry, departments of association; not only from the concerned country but also international bodies.

5.4 Digital Divide: Proposal

In successful Agricultural Informatics practice, Digital Divide is really important concern. This problem is very much pertinent in the developing countries. However, in developed countries also in certain areas, contexts, communities and localities Digital Divide can be considered as important concern and need to resolved by the proper initiative, technological gap fulfillment, training of the common people, people involved with the Agricultural Systems and so on. Furthermore, for healthy Agricultural Informatics practice services providers also need to act properly.

5.5 Government Support: Proposal

Government of each and every country is engaged with social services and other allied activities. Various projects, initiatives are the core of the ministries and departments of Governments of every country. Regarding effective Agricultural Informatics practice a proper and healthy support is always welcome and required routine wise. The modification, evaluation of the existing policies, projects also need to re-implement from time to time.

5.6 Awareness and Literacy on Agro Informatics: Proposal

Proper and healthy awareness is also important to enact Agricultural Informatics properly. Different associations, organizations, universities and Higher Educational Institutes (HEIs) and others should engage in healthy and proper awareness programs, training programs, exhibitions and extension services, etc. Government and concerned Ministries also may hold proper steps and initiation on real Agricultural Informatics practice by different means. Here farmers can also come under basic literacy and computer literacy if needed.

5.7 Skill Development: Solutions

Agricultural Informatics as a field combing two major fields therefore skill development is very important and required for healthy ICT applications in Agriculture. Here basic skill development can be offered with certificate, diploma and short term training program for the farmers, and also Agro Organization personnel for basic marketing and other allied activities using IT and Computing. Skill Development drive can also be offered by the departments, training organizations, universities, even concern agro firms for enhancing ICT in Agriculture. Further, a basic mobile based tutorial can also be effective in this regard [25, 26, 33].

5.8 Environment, Locality and Awareness: Proposal

As Agricultural Informatics deals with Environmental issues especially by the computing and allied products therefore it is suggested to use only the computer, technologies and machines which are less harmful to the environment. Further,as in different communities and locality, there may be an unwillingness in Agricultural Informatics practice therefore proper awareness schemes must be initiated by different stakeholders.

5.9 Policies, Framework and Regulations: Proposal

It is worthy to note that proper policies, regulation and frameworks at par the requirement of the concerned countries, territories and requirements are a must. Agricultural Informatics as deals with interdisciplinary facets and professional practice attributes therefore study needs proper policies and framework and this can be possible or initiated by the concerned government, administration, universities and higher educational institutions, agricultural institutes and associations.

5.10 Existing and New Technological Integration: Proposal

The new technologies are important and need of the hour. Agricultural Informatics or Agro ICT is required proper integration of newer technologies with the existing. Therefore, timely and frequent checking on existing technologies and systems with the integration of new ones should be provided as per the need and requirement.

5.11 Technological Development and Cycling: Proposal

Technological development is very much important in the Agricultural Informatics practice. Hence the involvement of the latest technologies such as Data Analytics and Big Data Management, Virtualization and Cloud Systems, Internet of Things, and Web of Things, Advanced Converged Network, Usability Engineering and UXD etc. should be offered at par with the other developed countries if possible, by ensuring possible means. Cycling is also very important and expected [6, 17, 30].

6 Conclusion

Agricultural Informatics is fully technology driven with basics of Computing Technologies and all the components of the Information Technologies and emerging components Data Analytics and Big Data Management, Virtualization and Cloud Systems, Internet of Things, and Web of Things, Advanced Converged Network, Usability Engineering and UXD, Multimedia Systems and so on. Therefore, Agricultural Informatics can be able in bringing of latest technologies and which may come with issues and challenges as well. The issues viz. Technological implementation, development of technologies, environmental issues, financial issues, awareness and skill development can be resolved with the initiation of proper planning, interest, policy and framework formulation and their solid implementation throughout and whenever needed. As far as developing and developed countries are concerned Agro Informatics is very important and effective but in many context proper initiation and steps are missing in some context, and it is the need of the hour regarding healthy and effective smart agricultural system development with sustainability.

References

1. Abbasi, A.Z., Islam, N., Shaikh, Z.A.: A review of wireless sensors and networks' applications in agriculture. Comput. Stand. Interfaces **36**, 263–270 (2014)
2. Adão, T., et al.: Hyperspectral imaging: a review on UAV-based sensors, data processing and applications for agriculture and forestry. Remote Sens. **9**, 1110 (2017)
3. Adetunji, K.E., Joseph, M.K.: Development of a cloud-based monitoring system using 4duino: applications in agriculture. In: International Conference on Advances in Big Data, Computing and Data Communication Systems (icABCD), pp. 4849–4854. IEEE (2018)
4. Ahmad, T., Ahmad, S., Jamshed, M.: A knowledge based Indian agriculture: with cloud ERP arrangement. In: International Conference on Green Computing and Internet of Things (ICGCIoT), pp. 333–340. IEEE (2015)
5. Aubert, B.A., Schroeder, A., Grimaudo, J.: IT as enabler of sustainable farming: An empirical analysis of farmers' adoption decision of precision agriculture technology. Decis. Support Syst. **54**, 510–520 (2012)
6. Babu, S.M., Lakshmi, A.J., Rao, B.T.: A study on cloud based Internet of Things: CloudIoT. In 2015 Global Conference on Communication Technologies (GCCT), pp. 60–65. IEEE (2015)
7. Balamurugan, S., Divyabharathi, N., Jayashruthi, K., Bowiya, M., Shermy, R.P., Shanker, R.: Internet of agriculture: applying IoT to improve food and farming technology. Int. Res. J. Eng. Technol. **3**, 713–719 (2016)
8. Bauckhage, C., Kersting, K.: Data mining and pattern recognition in agriculture. KI-Künstliche Intelligenz **27**, 313–324 (2013)
9. Channe, H., Kothari, S., Kadam, D.: Multidisciplinary model for smart agriculture using internet-of-things (IoT), sensors, cloud-computing, mobile-computing & big-data analysis. Int. J. Comput. Technol. Appl. **6**, 374–382 (2015)
10. Gill, S.S., Chana, I., Buyya, R.: IoT based agriculture as a cloud and big data service: the beginning of digital India. J. Organ. End User Comput. **29**, 1–23 (2017)
11. Gómez-Chabla, R., Real-Avilés, K., Morán, C., Grijalva, P., Recalde, T.: IoT applications in agriculture: a systematic literature review. In: Valencia-García, R., Alcaraz-Mármol, G., del Cioppo-Morstadt, J., Vera-Lucio, N., Bucaram-Leverone, M. (eds.) CITAMA2019 2019. AISC, vol. 901, pp. 68–76. Springer, Cham (2019). https://doi.org/10.1007/978-3-030-107 28-4_8
12. Goraya, M.S., Kaur, H.: Cloud computing in agriculture. HCTL Open Int. J. Technol. Innov. Res. **16**, 2321–1814 (2015)
13. Guardo, E., Di Stefano, A., La Corte, A., Sapienza, M., Scatà, M.: A fog computing-based iot framework for precision agriculture. J. Internet Technol. **19**, 1401–1411 (2018)
14. Martin, H.: The bad news is that the digital access divide is here to stay: Domestically installed bandwidths among 172 countries for 1986–2014. Telecommun. Policy **40**, 567–581 (2016)
15. Holster, H.C., et al.: Current situation on data exchange in agriculture in the EU27 & Switzerland. agriXchange, pp. 1–15 (2012)

16. Kamble, S.S., Gunasekaran, A., Gawankar, S.A.: Achieving sustainable performance in a data-driven agriculture supply chain: a review for research and applications. Int. J. Prod. Econ. **219**, 179–194 (2020)
17. Kajol, R., & Akshay, K. K.:Automated Agricultural Field Analysis and Monitoring System Using IOT. International Journal of Information Engineering and Electronic Business. 11, 17 (2018)
18. Khattab, A., Abdelgawad, A., Yelmarthi, K.: Design and implementation of a cloud-based IoT scheme for precision agriculture. In: 28th International Conference on Microelectronics, pp. 201–204. IEEE (2016)
19. Liu, S., Guo, L., Webb, H., Ya, X., Chang, X.: Internet of Things monitoring system of modern eco-agriculture based on cloud computing. IEEE Access **7**, 37050–37058 (2019)
20. Manos, B., Polman, N., Viaggi, D.: Agricultural and environmental informatics, governance and management: Emerging research applications. In: Z. Andreopoulou (ed.). IGI Global, Pennsylvania (2011)
21. Milovanović, S.: The role and potential of information technology in agricultural improvement. Econ. Agric. **61**, 471–485 (2014)
22. Muangprathub, J., Boonnam, N., Kajornkasirat, S., Lekbangpong, N., Wanichsombat, A., Nillaor, P.: IoT and agriculture data analysis for smart farm. Comput. Electron. Agric. **156**, 467–474 (2019)
23. Na, A., Isaac, W.: Developing a human-centric agricultural model in the IoT environment. In: 2016 International Conference on Internet of Things and Applications, pp. 292–297. IEEE, (2016)
24. Nandyala, C.S., Kim, H.K.: Green IoT agriculture and healthcare application (GAHA). Int. J. Smart Home **10**, 289–300 (2016)
25. Nayyar, A., Puri, V.: Smart farming: IoT based smart sensors agriculture stick for live temperature and moisture monitoring using Arduino, cloud computing & solar technology. In: Proceedings of the International Conference on Communication and Computing Systems, pp. 2–11 (2016)
26. Ojha, T., Misra, S., Raghuwanshi, N.S.: Wireless sensor networks for agriculture: the state-of-the-art in practice and future challenges. Comput. Electron. Agric. **118**, 66–84 (2015)
27. Othman, M.F., Shazali, K.: Wireless sensor network applications: a study in environment monitoring system. Procedia Eng. **41**, 1204–1210 (2012)
28. Ozdogan, B., Gacar, A., Aktas, H.: Digital agriculture practices in the context of agriculture 4.0. J. Econ. Finan. Account. **4**, 186–193 (2017)
29. Prantosh Kumar, P., Ghosh, M., Chaterjee, D.: Information Systems & Networks (ISN): emphasizing Agricultural Information Networks with a case Study of AGRIS. Scholars J. Agric. Veterin. Sci. **1**, 38–41 (2014)
30. Prantosh Kumar, P.: Information and knowledge requirement for farming and agriculture domain. Int. J. Soft Comput. Bio Inform. **4**, 80–84 (2013)
31. Paul, P.K., et al.: Agricultural problems in India requiring solution through agricultural information systems: problems and prospects in developing countries. Int. J. Inform. Sci. Comput. **2**, 33–40 (2015)
32. Paul, P.K., Aithal, P., Sinha, R., Saavedra, R., Aremu, B.: Agro informatics with its various attributes and emergence: emphasizing potentiality as a specialization in agricultural sciences—a policy framework. IRA-Int. J. Appl. Sci. **14**, 34–44 (2019)
33. Paul, P.K, Bhuimali, A., Ripu Ranjan Sinha, R.R.S., Aithal, P.S., Kalishankar, T., Saavedra M,R.: Agricultural data science as a potential field and promoting agricultural activities and sustainable agriculture. Int. J. Inform. Sci. Comput. **7**, 49–62 (2020)
34. Paul, P.K., Ripu Ranjan Sinha, R.R.S., Aithal, P. S., Saavedra M,R., Aremu, P.S.B.: Agro informatics with reference to features, functions and emergence as a discipline in agricultural sciences—an analysis. Asian J. Inform. Sci. Technol. **10**, 41–50 (2020)

35. Paul, P.K.: Agricultural informatics and practices—the concerns in developing and developed countries. In: Choudhury, A., Biswas, A., Singh, T.P., Ghosh, S.K. (eds.) Smart Agriculture Automation Using Advanced Technologies: Data Analytics and Machine Learning, Cloud Architecture, Automation and IoT, pp. 207–228. Springer, Germany (2021)
36. Paul, P.K.: Agricultural Informatics vis-à-vis Internet of Things (IoT): the scenario, applications and academic aspects—international trend & indian possibilities. In: Choudhury, A., Biswas, A., Prateek, M., Chakrabarty, A. (eds.) Agricultural Informatics: Automation Using the IoT and Machine Learning, pp. 35–65. Wiley-Scrivener, USA (2021)
37. Rezník, T., Charvát, K., Lukas, V., Charvát Jr, K., Horáková, Š., Kepka, M.: Open data model for (precision) agriculture applications and agricultural pollution monitoring. In EnviroInfo and ICT for Sustainability, pp. 97–107, Atlantis Press, Netharlands (2015)
38. Teye, F., Holster, H., Pesonen, L., Horakova, S.: Current situation on data exchange in agriculture in EU27 and Switzerland. In: Mildorf, T., Charvat, C., (eds.) ICT for Agriculture, Rural Development and Environment, pp. 37–47, Czech Centre for Science and Society Wirelessinfo, Prague (2012)
39. TongKe, F.: Smart agriculture based on cloud computing and IOT. J. Converg. Inf. Technol. **8**, 210–216 (2013)
40. Tsekouropoulos, G., Andreopoulou, Z., Koliouska, C., Koutroumanidis, T., Batzios, C.: Internet functions in marketing: multicriteria ranking of agricultural SMEs websites in Greece. Agrárinformatika/J. Agric. Inform. **4**, 22 (2013)
41. Zamora-Izquierdo, M.A., Santa, J., Martínez, J.A., Martínez, V., Skarmeta, A.F.: Smart farming IoT platform based on edge and cloud computing. Biosys. Eng. **177**, 4 (2019)

A Guava Leaf Disease Identification Application

Nikhil James, Kunal Kumar Shriwastav, Shilpita Medhi, and Smriti Priya Medhi[✉]

Assam Don Bosco University, Azara 781017, India
sp.medhi26@gmail.com

Abstract. Identification of plant diseases and their early intervention has a great role to play in the horticulture industry. The traditional methods adopted by farmers' may be time-consuming, costly, and occasionally wrong. With the advent of machine learning, it is now possible to address this issue with lightening speed. In this paper, we attempt to present the best approach suitable for detecting plant leaf diseases specific to guava (*Psidium Guajava*) tropical plant. We offer a suitable deep convolution neural network (CNN)-based method for diagnosing guava leaf illnesses in order to achieve the diagnosed output. The effectiveness of treatments depends on accurate disease diagnosis. The use of image processing in place of manual or visual detection of Guava leaf diseases alleviates the challenges, time commitment, and inaccuracies that would be encountered.

Keywords: Computer Vision · Machine Learning · Plant Diseases · Deep Learning · Image Processing

1 Introduction

Numerous guava plant diseases are significant elements that lower the amount, quality, and economic impact of production while still providing outstanding nutrients and minerals for human health. Humanity can benefit from enhanced immunity, improved vision, better control of diabetes, treatment for diarrhea, and stimulation of cognitive function, in addition to better oral, heart, thyroid, and skin health. Plant and leaf diseases are difficult, sluggish, and inaccurate to diagnose manually. Therefore, the current work makes an effort to identify the leaf samples using image processing in accordance with the five plant disease classes taken into account, which are Viburnum Chindo, Algal Leaf Spot, Rust, Curl, and Powdery Mildew as described in Table 1. The first crucial step in conducting a successful analysis is thus to get images of fruits and vegetables. Qualitative characteristics like colour and texture has to be also considered. These features from the sensor in the fruit picture collection are also influenced by illumination. Large-scale automatic fruit and vegetable monitoring may result from disease detection for fruits and vegetables using computer vision. This makes it easier to address specific risks before they affect real yields, such as the requirement to apply fertilizers to speed up growth. For e.g. yellowing leaves fall off the trees early leading to poor growth. Image processing along with machine learning techniques can be applied to a diseased plant leaf picture to make a machine understand its disease class and accordingly provide solutions to the farmers and cultivators for their early intervention and eventually protect the plant from further damage.

P. Das et al. (Eds.): AMRIT 2023, CCIS 1954, pp. 61–69, 2024.
https://doi.org/10.1007/978-3-031-47221-3_6

Table 1. List of guava leaf diseases considered

Guava leaf disease	Images
1. Canker	
2. Munnification	
3. Rust	
4. Dot	
5. Healthy	

2 Literature Review

P. Perumala, Kandasamy Sellamuthub, K.Vanithac, and V.K. Manavalasundaramd et al. 2021 [8] to identify several guava plant species, color histogram equalization and the unsharp masking technique were used. Use of deep convolutional neural network-based data augmentation is part of this method (DCNN). There are now nine altered plant photos thanks to nine 360-degree viewpoints. Innovative categorization networks were then fed with these improved data. Before processing, the proposed approach was normalized. The experimental analysis included data on guava illness that was locally acquired in Pakistan. SqueezeNet, GoogleLeNet, and five different neural network topologies are combined in the proposed study to identify different guava plant species. ResNet-101, which had a classification accuracy of 97.74%, generated the greatest results, according to the experiment's findings.

Swarn Avinash Kumar, Talha Meraj, Hafiz Tayyab Rauf et al. 2021 [10] the use of color histogram equalization and the unsharp masking technique to identify several guava plant species. This technique makes use of data enhancement based on deep convolutional neural networks (DCNN). Nine views from 360 degrees were used to increase the number of altered plant images. Afterward, this improved data were fed into cutting-edge classification networks. The suggested method was normalized before processing. A locally acquired Pakistani dataset on guava disease was used for the experimental investigation. The proposed study uses SqueezeNet and GoogLeNet along with five neural network topologies to recognize various guava plant types. The results of the experiment showed that ResNet-101, with a classification accuracy of 97.74%, produced the best outcomes.

Ahmad Almadhor, Hafiz Tayyab Rauf 2, Muhammad Ikram Ullah Lali et al. 2021 [6] recommended technological solutions For crop fields to remain productive, diseases in plants must be automatically detected as soon as they manifest on the plants' leaves and fruit. A system powered by artificial intelligence (AI) is described in this paper to identify and categorize the most prevalent guava plant diseases. The suggested framework uses 4E colour difference image segmentation to separate the diseased areas. Additionally, colour (RGB, HSV) histogram and textural (LBP) features are used to generate rich, useful feature vectors. While disease recognition is carried out by using sophisticated machine learning classifiers, colour and textural data are combined to identify and achieve equivalent results compared to 8 individual channels (Fine KNN, Complex Tree, Boosted Tree, Bagged Tree, Cubic SVM). On a high-resolution (18 MP) image dataset of guava leaves and fruit, the suggested framework is assessed. With 95% accuracy in identifying four guava fruit diseases (Canker, Mummification, Dot, and Rust) against healthy fruit, the Bagged Tree classifier produced the best recognition results. The suggested approach might assist the farmers in preventing potential production loss by exercising early vigilance.

A S M Farhan Al Haque; Rubaiya Hafiz; Md. Azizul Hakim; G. M. Rasiqul Islam et al. 2020 [9] A system for the diagnosis of guava disease and the provision of remedial suggestions is proposed in a paper using convolutional neural networks (CNNs). They have gathered pictures of guava from several Bangladeshi districts that have been free of anthracnose, fruit rot, and canker as well as guava that have all three diseases. The third model, which had an accuracy of 95.61%, surpassed the other two in this study's application of three CNN models. Performance criteria such as precision, recall, and F1 score are assessed and proven to produce excellent results for rigorous experimentation.

Harshal Waghmare and Radha Kokare 2016 [7] Technologies for leaf texture pattern analysis and plant disease analysis are suggested. This work is based on a technique for identifying grape leaf disease. After context removal, the device is employed as the input for segmentation on a single plant leaf. A high-pass filter for the leaf is used to assess the image segmentation of the diseased component. It is possible to obtain a unique sectional leaf texture. An excellent texture model is offered by locally based fractal features that are nature invariant. Each distinct sickness would have a different feel. The extracted texture pattern is subsequently graded using a multiclass SVM. For the processing of automated DSS (Decision Support Systems) and farmers, readily available multiclass SVM implementations are developed to identify the diseases observed in grape plants. The high pass filter is used to observe the sick area of the leaf as the scheme segments and examines a single leaf. A segmented leaf texture function that is invariant locally in nature is then retrieved using fractals, and the strong texture module is then provided. The texture removed pattern is then designateded as an SVM designation for multiclass in groups of classes that are either healthy or unwell, depending on the case.

Downy mildew and black, and red are the two main, often occurring diseases that are the focus of the investigation. 9 The suggestion method accurately and promptly gives farmers advice from agricultural specialists (96.6%).

Jitesh P. Shah et al. 2016 provide an analysis of several image processing and machine learning techniques for recognizing ill plants from pictures. In addition to examining various approaches, the publication also discussed important concepts in machine learning and image processing for identifying and classifying plant diseases. The breadth of this investigation led to the inclusion of 19 document-based experiments on ailments affecting plants, including those that harm rice, fruit, and other crops. Scale, lack of groups, preprocessing, technological segmentation, styles, classifier accuracy, and other factors affecting the collection of image data were all taken into account. It also used the poll to do additional research, enhance the identification, and classify a variety of rice-related diseases.

3   Proposed Plan

Adequa P. Perumala, Kandasamy Sellamuthub, K.Vanithac, and V.K. Manavalasundaramd et al. 2021 suggests a philosophy for recognizing guava leaf maladies early and precisely using image preparation techniques and Support Vector Machine (SVM). The suggested framework includes the following steps: prepreparing an image, segmenting it, clustering it using k-means, and extracting data using a grey level co-occurrence matrix (GLCM). The SVM classifier is then used to classify the image. In contrast to the current framework, the suggested framework essentially detects plant leaf disease at an early stage and increases accuracy to 98.17%.We are first building a model to diagnose the condition using just an image of a guava leaf and propose treatment alternatives by identifying guava leaf diseases with deep learning classification techniques and machine learning models. To design a programme that displays to the user the ailment that has been identified and its associated therapies.

Methodology of the Model

Machine learning is essential to the automation of many different systems. With that goal in mind, the proposed framework was designed using machine learning approaches, and we are currently working to detect diseases from plant leaf photos. This process entails a number of processes, each of which can be outlined as follows.

Step 1: Pre-processing an image. The stage of the framework where we resize the photographs is really important.

Step 2: Create a model with CNN: Convolutional neural networks are used as the deep-learning technique more frequently for picture classification and image identification. The best benefit of employing deep learning algorithms is that deep learning automates the process of extracting image features. The structure of CNNs allows for the mapping of visual data (or 2D multidimensional data) to output variables (1dimensional data). Since it has been demonstrated to be so effective, they are prepared to utilize it for any type of data prediction, including image classification. As a starting point, we use VGG-16.

Step 3: Training data from the processed data. The training data will be derived from the data that has been processed.

Step 4: Testing Data: The testing data, which classifies the disease's health and determines whether it is healthy or not, will also be obtained from the processed data.

Step 5: Assess performance.

We have used two popular CNN architecture, namely ResNet50V2 and VGG19. Firstly we collect the dataset from Kaggle and then labeled the images. Secondly we preprocessed the images and imported the ResNet50V2 architecture from pre trained model and then imported the VGG19 model. Lastly we evaluate the model using F1 score, accuracy score and confusion matrix. As of now we get an accuracy of 86%.

3.1 Experimental Analysis

Here, firstly we are developing a model and after that we will develop an app that can detect the disease by just a picture of the guava leaf and also give back some remedies for it. The guava leaf disease we are going to detect are:

1. Canker
2. Dot
3. Mummification
4. Rust
5. Healthy (Table 2)

Table 2. Dataset Description

Disease Category	Train Sample	Test Sample	Valid Sample
Canker	198	62	49
Dot	198	62	49
Mummification	198	62	49
Rust	198	62	49
Healthy	198	62	49
Total	990(80%)	310(20%)	248(20%)

We have taken the dataset from Kaggle data. This dataset consists of 1550 image files of 5 different classes, namely canker, dot, mummification, Rust, healthy. We are in the process of detection of diseases from plant leaf images involves vari ous steps and each of those steps can be discussed as follows -

Step 1: Image Pre-processing

Step 2: Model design using CNN

Step 3: The training data will be obtained from the processed data.

Step 4: The testing data will also be obtained from the processed data, and it classifies the disease.

Step 5: Evaluate performance

Metrics involves in evaluating the performance of the model are as follows:

Confusion Matrix

A confusion matrix, also known as an error matrix, in the field of machine learning and specifically the issue of statistical classification, is a particular table structure that enables visualization of the performance of an algorithm, typically a supervised learning one (in unsupervised learning it is usually called a matching matrix). Both variations are found in the literature. In the matrix, each row represents the cases in an actual class, whereas each column represents the instances in a forecast class (Fig. 1).

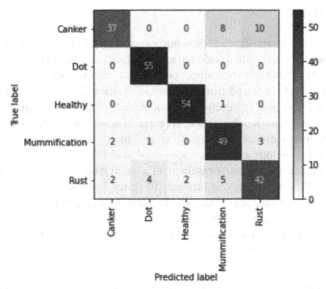

Fig. 1. Confusion Matrix

Accuracy Score: One parameter for assessing classification models is accuracy. The percentage of predictions that our model correctly predicted is known as accuracy. Accuracy is defined as follows in formal language.

$$Accuracy = \frac{Number\ of\ correct\ predictions}{Total\ number\ of\ predictions} \tag{1}$$

For binary classification, accuracy can also be calculated in terms of positives and negatives as follows:

$$Accuracy = \frac{TP + TN}{TP + TN + FP + FN} \tag{2}$$

where TP = True Positives, TN = True Negatives, FP = False Positives, and FN = False Negatives

F1 score: By calculating their harmonic means, the F1-score integrates a classifier's precision and recall into a single metric. Comparing the effectiveness of two classifiers is its main purpose. Assume classifier B has a better precision and classifier A has a larger recall. In this situation, it is possible to tell which classifier yields superior results by comparing their F1 scores. The F1-score of a classification model is calculated as follows:

$$\frac{2(P * R}{P + R} \tag{3}$$

P = the precision
R = the recall of the classification model

4   Future Work

Due to its useful bio composites and previously mentioned traditional uses that promote health, guava is a product with a wide range of applications. The existence of potential plant compounds of interest to benefit people generates an exciting area of study, which is constantly driven by demand from real human systems for the development of circular economies. The goal of developing value-added goods or techniques that exploit underused biomass sources is a high-impact topic in several industries that will help to address numerous nutritional, cosmetic, food, and environmental problems. This study can be expanded to include a variety of additional guava diseases, including expanding the dataset. To achieve even better results with a much larger dataset, transfer learning techniques like MobileNet etc. can be used. To better differentiate between plant diseases, we will grow our database. This expert system is thought to be more user-friendly, practical, and the foundation for future ones.

It is intended to be made more accessible to users at all times and from any location by adding more guava disorders.

5   Conclusion

Compared to traditional diagnoses, farmers can acquire the diagnosis quicker and more precisely. This knowledge-based system offers an intuitive user interface and does not require much training to utilize it. Many tasks can be completed through deep learning more quickly and effectively. The ability to categorize images has improved thanks to deep learning and model training on KAGGLE data sets. Difficult tasks, such as disease identification in guava leaves, formerly required experts and a lot of effort. Users will be able to identify the ailment in its earliest stages with the aid of this Android application. We are working on the model and conducting a literature review in this phase, and in different phase we will be implementing the performance evaluation in the model as well as the application part. The proposed model achieved an accuracy of 84%. To manage guava leaf diseases, it is important to implement several suggestive measures . in guava leaf detection in machine learning the process involves collecting a diverse dataset of guava leaf images preprocessing ,feature extraction and model training , the model is trained and validated to detect guava leaf in new images

References

1. Mukherjee, M., Pal, T., Samanta, D.: Damaged paddy leaf detection using image processing. J. Global Res. Comput. Sci. **3**(10), 07–10 (2012)
2. Al-Hiary, H., Bani-Ahmad, S., Reyalat, M., Braik, M., Alrahamneh, Z.: Fast and accurate detection and classification of plant diseases. Int. J. Comput. Appl. **17**(1), 31–38 (2011)
3. Wahid, K., Bhowmik, P., Dinh, A., Islam, M.: picture segmentation and a multiclass support vector machine are used to detect potato illnesses. In: IEEE 30th Canadian Conference on Electrical and Computer Engineering, pp. 1–4. IEEE (2017)
4. Es-Saady, Y., El Massi, I., El Yassa, M., Mamma, D., Benazoun, A.: Based on the serial combination of two SVM classifiers, automatic disease recognition of plant leaves. In: 2016 International Conference on Electrical and Information Technologies (ICEIT), pp. 561–566. IEEE (2016)

5. Mainkar, P.M., Ghorpade, S., Adawadkar, M.: Plant leaf disease detection and classification using image processing techniques. Int. J. Innov. Emerg. Res. Eng. **2**(4) (2015)
6. Alharbi, A., Alouffi, B., Damaevius, R., Rauf, H.T., Almadhor, A., Lali, M.I.U., Rauf, H.T.: AI-Driven Framework for Machine Learning-Based Disease Recognition in Guava Plants Using High Resolution Images from DSLR Camera Sensors the sensors 2021, 21 and 3830 (2021)
7. Waghmare, H., Kokare, R., Dandawate, Y.: Detection and classification of illnesses of Grape plant using opposite colour Local Binary Pattern feature and machine learning for automated Decision Support System. In: 3rd International Conference on Signal Processing Integrated Networks, pp. 513–518 (2016)
8. Perumala, P., Sellamuthub, K., Vithac, K., Manavalasundaramd, V.K.: Classification of Guava Leaf Disease Using SVM, vol. 12, no. 7, pp. 1177–1183 (2021)
9. Hafiz, R., Hakim, M.A., Islam, G.R.: A computer vision system can recognise the illness in guavas and provide a cure using a deep learning approach. In: The Proceedings of the 22nd International Conference on Computer and Information Technology (ICCIT), pp. 1–6. Dhaka, Bangladesh, December 18–20, 2019
10. Avinash Kumar, S.: Deep convulational Neural Network. Appl. Sci. **12**(1), 239 (2022). Guava leaf disease according to Talha Meraj's classification
11. Tai, A.P., Martin, M.V., Heald, C.L.: Future climate change and ozone air pollution threats to global food security. Nat. Clim. Chang. **4**, 817–821 (2014). https://doi.org/10.1038/nclimate2317
12. Fergus, R., Dr. Zeiler: In Computer Vision-ECCV 2014. In: Fleet, D., Pajdla, T., Schiele, B., Tuytelaars T.: Visualizing and Understanding Convolutional Networks, pp. 818–833. Springer, Cham (2014). https://doi.org/10.1007/978-3-319-10593-2
13. Schmidhuber, J.: Neural networks for deep learning: an overview. Neural Netw. **61**, 85–117 (2015). https://doi.org/10.1016/j.neunet.2014.09.003
14. Sankaran, S., Mishra, A., Maja, Ehsani, R.: detection of huanglongbing in citrus groves using visible-near infrared spectroscopy. Comput. Electron. Agric. **77**, 127–134 (2011). https://doi.org/10.1016/j.compag.2011.03.004

Text to Image Generation Using Attentional Generative Adversarial Network

Supriya B. Rao[1], Shailesh S. Shetty[2], Chandra Singh[3][(✉)], Srinivas P. M[4], and Anush Bekal[3]

[1] Department of Artificial Intelligence and Machine Learning, New Horizon College of Engineering Bangalore, Mangaluru, Karnataka, India
[2] Department of Computer Science and Business System, Srinivas Institute of Technology, Mangaluru, Karnataka, India
[3] Department of Electronics & Communication Engineering, Sahyadri College of Engineering & Management, Mangaluru, Karnataka, India
chandrasingh146@gmail.com
[4] Department of Computer Science Engineering, Sahyadri College of Engineering & Management, Mangaluru, Karnataka, India

Abstract. AttnGan – Attentional Generative Adversarial Network (GAN) is a type of GANs that is used for text to image generation. Unlike other art models, it focuses on fine-grained word level information and once the model is trained by mapping the images and the corresponding text in the dataset, the model will generate image part-by-part using the image description given by the user as an input to the model. Current paper, focuses on the conversion of text into images based on the description provided by the user. This enables them to translate their ideas in the form of text into images of their required choices. The model is initially trained by the description of the text and images and when the user provides the text as an input to the system, it is used as a testing data. By analyzing the text word-by-word, the respective images pertaining to the text is generated part-by-part by considering the text description into account. This can heavily help the users in a applications such as in editing aids, documental archives, etc.

1 Introduction

Images are basically the representation of external form of a person or thing in the form of an art or in an artistic way. Images are a part of our daily lives. We see some beautiful sceneries around us in our daily life or may be while travelling and capture images of them [1, 2]. There are many people who are artistic by nature. They try to recreate the situation or the experience by trying to generate the same scenery in the form of an art. Art maybe of many types. Some artists prefer depicting their ideas in the form of a cartoon art, and some prefer generating landscape arts which looks very much realistic [3, 4]. But some common people or average designers will be having ideas but are not able to generate realistic images like other artists. So, our project will be aiming to provide assistance to such people and designers by generating realistic images using simple natural language text using Attentional Generative Adversarial Networks. There is good

P. Das et al. (Eds.): AMRIT 2023, CCIS 1954, pp. 70–82, 2024.
https://doi.org/10.1007/978-3-031-47221-3_7

demand for system generated realistic images in various industries. For example, some of the industries like photograph industry would use it to generate realistic photographs as such. Or they might also use it for improving the resolution of a given image [5, 6]. The 3D modelling industries can also make use of it. They can generate realistic human faces using this concept. People would always prefer an easier way to get things done. And when a system can generate realistic images on its own, just based on visual descriptions, the market would have a great demand for it. The main purpose of this paper is to create a model which can help many industries like Art industry, Modelling industry, Photography industry and many such industries and also normal people by generating realistic images through the system based on the descriptions provided by them about the image [7, 8]. This enables the former and the latter to truly achieve their potential and goal and also save time even though they are unable to portray their innate talents on a sheet of paper or as precised as a machine does. The concept of the paper can be further extended to support a huge range of artists including authors, musicians where the cover art can be generated based on the artists" likes and dislikes. And this can also assist other image and modelling industries to get better resolution images and 3D modelling designs. This can majorly aid them to lessen the burden that is required to think of an idea and then create them on the concerned medium. Since the 3D modelling and gaming industries, art and photo industries are all getting bigger day by day, it became very important to create a system which can generate realistic images based on textual description given about the image. New concepts like Generative Adversarial Network became widely accepted for synthesizing images and videos. This model will allow customer to get realistic system generated images as per what they desired. And it will reduce the time consumption factor in an efficient manner to do the same if itwas to be done manually [9, 10].

2 Problem Statement

Generative Adversarial Networks have gained prominent importance in the world today wherein they are widely employed in a multitude of applications. The Problem Statement for our project is Text to Image Generation with Attentional Generative Adversarial Networks (AttnGAN). The AttnGAN can synthesize fine-grained details at different subregions of the image bypaying attentions to the relevant words in the natural language description. In addition,a deep attentional multimodal similarity model is proposed to compute a fine-grained image-text matching loss for training the generator.

3 System Design

An Architecture diagram, basically represents the overall working of the model by depicting different components in the project. So here initially, we get the data from the dataset which consists of images and captions pertaining to these images [11, 12]. The data willbe given to the skip thought model during preprocessing. The job of skip thought modelis that it encodes the sentences and extracts the semantic meaning of the it and try to understand the context of it.

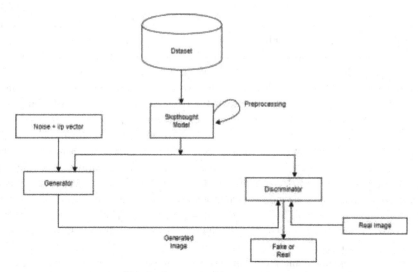

Fig. 1. System Architecture Diagram

As shown in Fig 1, now this encoded data will be given to Generator and Discriminator. Also, a random input will be given to the generator, which is our input text. Now the generator network converts this into data instance and creates an image. This will be sent to discriminator and the discriminator network will classify the generated data. Now the discriminator gives an output saying the image is real or fake by analyzing samples of real world images. If the output turns out to be real then the image is accepted, else this will be considered as a feedback by generator and will repeat the same process [13, 14].

A Flow chart is a diagrammatic representation of workflow of a process. It diagrammatically represents the algorithm used in the process. It is basically step by step approach towards solving a task. In this diagram, the input text at the beginning which is given to the text embedding model. This gives, out as the embedded text. Now this and some noise vector samples are together considered as input vector and this will be the input for the generator network which generates the image based on the input [15, 15]. Now the discriminator decides if the generated image is real or fake. And based on this, loss optimization functions are updated on both generator and discriminator. Once the generated image convinces the discriminator that it is real, that image will be displayed to the user via interface.

Figure 2 shows flow chart pertaining to step-by-step generation of a realistic image from a text input. A data flow diagram is basically a way of representing how the data flows in a system or a process. It provides a graphical representation of the hierarchy in which the data flows in a system. Images of different flowers and batch of captions for each of those images is basically what our dataset consists of. The text describing what we want to generate, which is typed by the user is considered as input to the system. That is given to the Skip thought model which is a part of preprocessing, to analyze the semantic properties of the input text. Now a generator is being defined which is basically responsible for image generation of images. Also discriminator is being defined which is

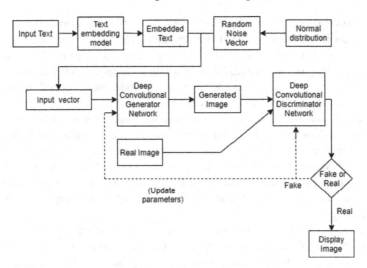

Fig. 2: Flow diagram of proposed text to image generation.

responsible for classifying if the generated image is real or fake. Both the generator and discriminator accept different parameters for models such as optimizers, loss functions etc. The data now passes through the generator which converts it into a data instance. And this in turn will be moving towards the discriminator which gives its output [15, 16]. Now a network is set up between generator and discriminator so that the feedback from the discriminator is considered by the generator and will eventually improve until it generates an image thatis real enough for the discriminator approves it.

Now every time the generator fails to convince the discriminator that the generated image is real, we will have a loss and inorder to minimize the loss we have defined a loss optimizer function. So we train the model with this setup and finally arrive at the result which is a realistic image generated as per the input text [17].

4 Implementation

The first step is to obtain the dataset for which we have used the dataset created by Caltech-UCSD Birds-200-2011 which is publicly available online. The text data which describes each image in the dataset is also obtained from a publicly available dataset.

Firstly we have to create a one hot encoding for the textual data that is present in the input. We split the text based on newline character and create an array on $n \times 1$ dimension initialized with zeros where n is the size of the individual text data. We use a library to perform the encoding whose output is an array of numbers. Example: input:" w i n d o w " output: [0 0 0 0 0 0 1 0 0 0 0 0]. Then we have to save the array as a vector along with the image caption and image class. We then have encoded the captions using skip thought model present in a library. The encoded captions are stored in separate files using the format of pickle „pkl". All of the above functions fall into data pre-processing step. The obtained pickle files are used for training the GAN model.

First we build an encoder model called Neural Network Image Compression Encoder. Compresses an image to a binarized NumPy array. The image must be padded to a multiple of 64 pixels in height and width. To build the encoder we need the input image which can be of jpeg or png format, iteration which is the quality level for encoding the image, directory to save the output encoding. The encoder model is built using Tensor Flow. The model used for compressing is called residual gru.pb.

Next, we created an encoder model for text data. This encoder model is to take the text as input which describes the image which has to be generated after which we have to load the skip thought model to generate the encoded vector. Next we have to build the decoder model Neural Network Image Compression Decoder to decompress an image from the numpy"s npz format generated by the encoder. This takes input of the binary coded file produced by the encoder, number or iterations and the output directory path to save the image.

Then we have to write code for both generator and discriminator network for which we have used Deep Convolution Neural Network to achieve this step. Encoding of texts and random noise will be taken as input by the generator. The discriminator will keep classifying the image as fake until the generator is able to produce an image that can be classified as a real image. Therefore both the generator and discriminator keep improving by iteratively getting updated with the help of loss optimizer function. Then we trainour model with respect to our dataset. Once the training process is completed, we can continue with the testing process. Finally we will get a model that generates a realistic image when the user gives an image description as an input caption to the model.

CNN or Convolution Neural Network is an algorithm that falls under the category of Deep Learning or Deep Neural networks. This is most likely used for analyzing the visual imagery, image processing etc. This is quite useful when compared to other image classification algorithms as it uses minimal processing. It has many advantages in comparison to other similar algorithms as it is independent of human intervention.

CNN has 4 basic steps, i.e., Convolution layer, Activation layer, Pooling layer and fully connected layer. In the convolution layer, different features or filters are considered. A kernel of size 3×3 or 5×5 is considered and is moved throughout the image matrix. We then get matrices with certain values after taking the mean of all the multiplied pixel values while the kernel is moved throughout the matrix for each filter as such. Then we apply activation layer to these obtained convolution matrices. An activation layer is applied so that the values in the matrix falls in a given particular range so that it becomes easy for later computations.

Then we apply pooling layer over these matrices. Here we again take another window of a smaller size and move it throughout the matrices of each of the filters. Here, we select the maximum value among the values in the window currently. So eventually we get smaller matrices and these depict the pixel that matters the most for those respective features or filters. Then finally we have fully connected layer. All the values in the matrices obtained after applying pooling layer is stacked and using this the algorithm does image classification.

I Gradient Descent

Almost all Deep Learning algorithms use the Gradient Descent concept. A gradient is actually a vector, whose components consists of partial derivative of a function and the points of the same function that is moving in the increasing direction. A Gradient Descent is basically used to optimize the algorithms through minimizing some functions by moving in its steepest negative direction of its gradient. In the case of our project, we use it during loss optimization process. Here we use something called as a Gradient Tape. Using learning rates and few inputs given to the Discriminator, the GAN models are iteratively updated by this loss optimization function which will eventually help the model to generate a realistic image as an output.

II Training the Model

We take 3 inputs, sr, sw and sf into consideration and also another parameter α which is the learning rate. We then use the concept of Gradient Descent in order to optimize thealgorithm. This loss optimizer keeps iteratively updating the Generator and Discriminator models. We use a pre- trained Google News model for training purpose. Then finally we train our model with respect to our dataset.

III System Testing

Testing is the process of detecting any sort of error in implementation. This allows usto verify if the product has met the requirements that are specified in system requirement specification. Testing helps to improve the product quality. Programming testing is the way of checking a framework with the intention behind recognizing any errors, gaps or missing prerequisite against the genuine necessity. Programming testing is exclusively divided into two types – functional testing and non-functional testing.

Table 1: Test case 1

TC#	Caption	Expected Result	Actual Result	Status
TC-1	yellow least fly catch	a yellowish bird	a yellowish bird	Pass

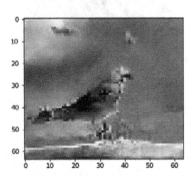

Fig. 3. Image for Test Case 1

Table 2: Test case 2

TC#	Caption	Expected Result	Actual Result	Status
TC-2	brown black feathers long neckblack head long black bill testing report	a brownish black birdwith long neck and beak	a brownish black birdwith long neck and beak	Pass

Fig. 4. Image for Test Case 2

Table 3: Test case 3

TC#	Caption	Expected Result	Actual Result	Status
TC-3	brown head and body with a light brownish tan stomach testing report	a brownish bird with a tannish stomach	a small bird with a tannish stomach	Pass

Fig. 5. Image for Test Case 3

Table 4: Test case 4

TC#	Caption	Expected Result	Actual Result	Status
TC-4	bird has a black beak with a long black neck testing report	a blackish bird with a beak and long neck	a black bird with black beak	Pass

Fig. 6. Image for Test Case 4

Table 5: Test case 5

TC#	Caption	Expected Result	Actual Result	Status
TC-5	bright yellow bird with orange beak testing report	a yellowish bird with orange beak	a yellowish orange bird	Pass

Fig. 7. Image for Test Case 5

Table 6: Test case 6

TC#	Caption	Expected Result	Actual Result	Status
TC-6	small bird with royal blue crown black eye baby blue colored billtesting report	a small blue bird with bluish beak	a bluish bird	Pass

Fig. 8. Image for Test Case 6

5   Results and Discussion

As a result, we can generate images based on CUB dataset, but the images do not look realistic enough. However, GAN-INT and GAN-INT-CLS show commendable images that usually matches all or at least a part of the caption. By going through and analysing different literature surveys, we conclude that it is quite hard to generate a bird image, as birds have stronger structural regularities across species that make it easier for Discriminator to spot a fake bird image.

```
def random_flip(image):
  image = tf.image.flip_left_right(image)
  return image.numpy()

def random_jitter(image):

  image = expand_dims(image, 0) #add additional dimension necessary for zooming
  image = image_augmentation_generator.flow(image, batch_size=1)
  result = image[0].reshape(image[0].shape[1:]) #remove additional dimension (1, 64, 64, 3) to (64, 64, 3)
  return result
```

Fig. 9. Code to invert and zoom the image

Pose and Background Style Transfer and Sentence Interpolation The images generated from our paper as a result of processing a text description appears to be reasonably realistic. Since we keep the noise distribution the same, the only changing factor within each row is the text embedding that we use. Interpolations can accurately reflect color information.

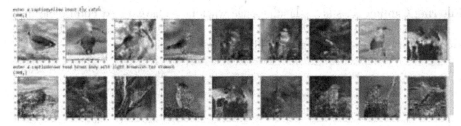

Fig. 10. Background information displayed due to style transfer

```
def create_sentence_embeddings():
    df = pd.read_csv('final.csv')
    #path=PurePath('./word2vec_pretrained_model').resolve().parent()
    model=gensim.models.KeyedVectors.load_word2vec_format('C:\\Users\\Dell\\Desktop\\final year\\text-to-image-synthesis\\word2vec_pretrained_model\\GoogleNews-vectors-
negative300.bin',binary=True)
    cleaned_captions=clean_and_tokenize_comments_for_image(df['captions'].values)
    cleaned_images=df['images'].values
    print("Tokenizing completed")
    n1,n2=create_feature_vectors_for_single_comment(model,cleaned_captions,cleaned_images)
    word_vector_dict=dict(zip(n1,n2))
    pickle.dump(word_vector_dict,open('word_vector_min_bird'+'.p','wb'))
```

Fig. 11. Code for creating text embedding

```
discriminator = layers.Conv2D(filters=64, kernel_size=(3, 3), padding="same")(merge)
discriminator = layers.LeakyReLU(0.2)(discriminator)
discriminator = layers.GaussianNoise(0.2)(discriminator)
```

Fig. 12. Code for noise distribution

Front End We have built an app which works on Flutter framework. Initially we have an opening page which is called as a splash screen which will then route the user to the main landing page.

TEXT2IMAGE

Text to image generation app

Fig. 13. Splash Screen

Entering Caption This is the landing page wherein we have a Text Form Field that is meant for entering caption. This will be taken as input by the model which will be processed by our GAN model and it will return the generated images.

Fig. 14. Entering caption

Once all the processing is done, the model will return 9 images that are being generated by our GAN model. We are displaying the generated images in the app in list view as shown in Fig 15.

Fig. 15. Displaying the generated images

6   Conclusion and Future Work

In this paper, we have developed a simple and effective model to generate realistic images of birds based on detailed visual descriptions. We demonstrated that the model can generate many decent and reasonable visual interpretations of a given text caption. Our manifold interpolation regularizer considerably improved the text to image generation on Caltech-UCSD Birds-200-2011 dataset. We showed non mix up of style and content, and bird pose and background transfer from query images onto text descriptions. In future work, we aim to further scale up the model to higher resolution images and train with wider range of datasets so that we can generate images with respect to many more textthat can be a description about any sort of general object or creature or description abouta scene for that matter.

References

1. Xu, T., Zhang, P., Huang, Q., Zhang, H., Gan, Z., Huang, X., He, X.: Attngan: fine- grained text to image generation with attentional generative adversarial networks, In: Proceedings of the IEEE Conference on Computer Vision and Pattern Recognition, pp. 1316–1324, (2018)
2. Zhang, H., Xu, T., Li, H., Zhang, S., Wang, X., Huang, X., Metaxas, D.N.: Stackgan: Text to photo-realistic image synthesis with stacked generative adversarial networks, In: Proceedings of the IEEE International Conference on Computer Vision, pp. 5907– 5915, (2017)
3. Zhu, J.Y., Park, T., Isola, P., Efros, A. A.: Unpaired image-to-image translation using cycle-consistent adversarial networks, In Proceedings of the IEEE International Conference on Computer Vision, pp. 2223–2232, (2017)
4. Reed, S., Akata, Z., Yan, X,. Logeswaran, L., Schiele, B., Lee, H.: Generative ad- versarial text to image synthesis, arXiv preprint arXiv:1605.05396, (2016)
5. Mansimov, E. Parisotto, E. Ba, J L., Salakhutdinov, R.: Generating images from captions with attention, arXiv preprint arXiv:1511.02793, (2015)
6. Qiao, T., Zhang, J., Xu, D., Tao, D.: Mirrorgan: Learning text-to-image generation by redescription.(2019)," arXiv preprint arXiv:1903.05854, (2019)
7. Creswell, A., White, T., Dumoulin, V., Arulkumaran, K., Sengupta, B., Bharath, A.A.: Generative adversarial networks: an overview. IEEE Signal Process. Mag. **35**(1), 53–65 (2018)
8. Karras, T., Laine, S., Aila, T.: A style-based generator architecture for generative adver- sarial networks, In: Proceedings of the IEEE Conference on Computer Visionand Pattern Recognition, pp. 4401–4410, 2019
9. Sharma, S., Suhubdy, D., Michalski, V., Kahou, S.E., Bengio, Y.: Chatpainter: Improving text to image generation using dialogue, arXiv preprint arXiv:1802.08216, (2018)
10. L. Wang, V. Sindagi, and V. Patel, "High-quality facial photo-sketch synthesis usingmulti-adversarial networks, In: 2018 13th IEEE International Conference on Automaticface & Gesture Recognition (FG 2018), pp. 83–90, IEEE, (2018)
11. Wang, X., Tang, X.: Face photo-sketch synthesis and recognition. IEEE Trans- actions on Pattern Anal. Mach. Intell. **31**(11), 1955–1967 (2008)
12. Goodfellow, I. et al: Generative adversarial nets, In Advances in neural information processing systems, pp. 2672–2680, (2014)
13. Mirza, V., Osindero, S.: Conditional generative adversarial nets, arXiv preprint arXiv:1411.1784, (2014)
14. Salimans, T., Goodfellow, I., Zaremba, W., Cheung, V., Radford, A., Chen, X.: Improved techniques for training gans, in Advances in neural information processingsystems, pp. 2234–2242, (2016)
15. Desai, X., Shetty, A.D.: Electrodermal activity (EDA) for treatment of neurological and psychiatric disorder patients: A Review," 2021 7th International Conference on Advanced Computing and Communication Systems (ICACCS), pp. 1424–1430, doi: https://doi.org/10.1109/ICACCS51430.2021.9441808 (2021)

16. Gurupur, V.P., Kulkarni, S.A., Liu, X., Desai, U., Nasir, A.: Analysing the power of deep learning techniques over the traditional methods using medicareutilisation and provider data. J. Exp. Theor. Artif. Intell. **31**(1), 99–115 (2019). https://doi.org/10.1080/0952813X.2018. 1518999
17. Shetty, Shruthi H., et al.: Supervised machine learning: algorithms and applications. fundamentals and methods of machine and deep learning: Algorithms, Tools and Applications: 1–16 (2022)
18. Singh, Chandra, K. V. S. S. S. S. Sairam, and Harish MB.: Global fairness model estimation implementation in logical layer by using optical network survivability techniques. International Conference on Intelligent Data Communication Technologies and Internet of Things. Springer, Cham, (2018)
19. Rahul Vijay, S., et al.: Object tracking robot using adaptive color thresholding.In: 2017 2nd International Conference on Communication and Electronics Systems (ICCES). IEEE, (2017)
20. Kanduri, S., et al.: Broadband services implementation by using survivable ATM architecture. Proceedings of International Conference on Sustainable Computing in Science, Technology and Management (SUSCOM), Amity University Rajasthan, Jaipur-India. 2019
21. Sairam, Kanduri, et al.: Ring architecture analysis implementation by using fiber network survivability techniques.In: Proceedings of International Conference on Sustainable Computing in Science, Technology and Management (SUSCOM), Amity University Rajasthan, Jaipur-India. (2019)

Attention-CoviNet: A Deep-Learning Approach to Classify Covid-19 Using Chest X-Rays

Thejas Karkera[1], Chandra Singh[1(✉)], Anush Bekal[1], Shailesh Shetty[2], and P. Prajwal[3]

[1] Department of Electronics Communication Engineering, Sahyadri College of Engineering and Management, Mangalore, India
chandrasingh146@gmail.com
[2] Department of Computer Science and Business System, Srinivas Institute of Technology, Mangaluru, Karnataka, India
[3] Department of Computer Science (AI/ML), Sahyadri College of Engineering and Management, Mangalore, India

Abstract. Covid-19 pandemic is spreading across the world at a breakneck pace. As of September 2022, JHU CSSE COVID-19 Data and Our World in Data have recorded up to 5.2 M deaths due to Covid-19. One of the essential steps of fighting this pandemic is to detect the disease early. Studies show some abnormalities in the chest X-rays of the Covid-19 patients, and these features help classify the covid-19 positive patients from the negative patients. In recent years in the field of medical x-rays, Deep Learning models have proven to have the ability to learn and identify features that a trained person can only identify. Due to the exponential increase in the covid-19 cases, the waiting time for covid results has increased, resulting in the late diagnosis of patients for Covid-19. Classifying a patient as Covid-19 positive for a genuinely covid positive patient is as essential as classifying a patient Covid-19 negative who is genuinely covid-negative. Here we present a detailed quantitative analysis on the performance of state-of-the-art models like ResNet-50, Dense-Net, Mobile-Net-V2, and MNAS-Net on classifying patients for Covid positive, Pneumonia positive, and Normal by evaluating Accuracy, Precision, Recall, and Jaccard Index. We propose a new deep neural network classification model for low-end devices that uses two Attention mechanisms. Figure 1. shows the attention map with confidence percentage for each classes. Our model uses fewer parameters and FLOPS than other state-of-the-art models and recorded a 2% increase in Accuracy and other evaluation parameters. The dataset used for implementation is a public dataset COVID-DATASET and NIH Chest X-ray from kaggle. It has over 1000 images of Covid Chest X-ray, Pneumothrax, Mass, Pneumonia, Cardiomegaly, Nodule, Effusion, Atelectasis and inflitration. All the model implementations are implemented on PyTorch.

Keywords: COVID-19 · Deep Learning · Attention · Classification · Covid Chest · X-ray · Pneumonia

P. Das et al. (Eds.): AMRIT 2023, CCIS 1954, pp. 83–99, 2024.
https://doi.org/10.1007/978-3-031-47221-3_8

1   Introduction

Since the first time coronavirus got detected in Wuhan, China, in December 2019, there have been more than 10million confirmed cases all over the world. After a year, there have been more than 3million confirmed casualties due to coronavirus [1, 2]. Reverse-transcription-polymerase-chain-reaction (RT-PCR) is found as a method for COVID-19 screening. Due to a shortage in the testing kits for RT-PCR in recent months, early diagnosis of covid cases is essential. There is an observation that the positive rating of RT-PCR for swab samples from the throat is 30% to 60% which may also infect a large population of healthy patients. CT scans of chest x-rays have shown very high sensitivity for covid-positive patients. X-ray images of covid patients have also shown covid presence visually [3–5]. In recent coronavirus research, patterns resembling ground glass are observed in the edges of pulmonary vessels, challenging to examine visually in the early stages of COVID-19. The Covid-19 reports also reported Asymmetric patchy opacities. Expert radiologists can only find these abnormalities. Let's consider the massive rate of affected people and a limited number of skilled radiologists. Automatization of detecting such abnormalities can be a considerable breakthrough for assisting in the diagnosis procedure and increasing the rate of early detection of Covid-19 [6]. Deep Convolutional Neural networks have proven to detect these features with a high accuracy rate. In recent research, traditional machine learning techniques are used to detect Covid-19 disease in patients from chest X-rays [7]. Machine learning. Techniques usually involve user-defined or already existing feature extraction procedures and recognition tasks. Deep Convolutional Neural Networks (DCNN) have better performance than conventional machine learning models. Deep Learning models are used for face recognition to high-resolution aerial image segmentation and classifications. DCNN has better analysis properties in the medical field, which includes segmentation, classification, and detection [8]. An early breakthrough in Convolutional networks came in the form of AlexNet; it introduced building blocks like max-pooling, convolutions, and dense blocks. To reduce over-fitting global-average pooling layers and a new regularization method was used [9]. Vgg-16 showed improvements in classification tasks by increasing the network depth by using small filters of size 3 × 3. GoogleNet's primary objective in 2014 was to reduce the complexity involved in computations by incorporating Inception Layers. In 2016, an architecture with residual connection was developed for deeper convolutional networks to tackle the loss in feature information in the later stage of the network. A Dense-Net architecture highlighted the importance of feature reuse from the previous layers in. DenseNets was able to obtain improvements on then state-of-the-art whilst keeping the memory requirement less and achieved a computationally high-performance measure [9–11]. In recent years, a requirement for models with fewer parameters came into the picture. Mobile-Net V2 was able to achieve state-of-the-art performance by incorporating inverted-residual blocks and linear bottleneck layers into the architecture. Mnas-Net tackled the need for the creation of a platform-aware network for mobiles by using fewer FLOPS than Mobile-Net and achieving 1.8x faster performance by incorporating a factorized search method [12, 13]. Here the authors present a detailed quantitative and qualitative analysis on Deep Learning on chest x-ray images for Covid-19 detection. We train five deep learning models from each year VGG-9, ResNet50, DenseNet-121, MobileNet-V2, and MNAS-Net. MobileNet-V2 and MNAS-Net have performed very well for various image classification tasks with fewer parameters than other state-of-the-art models. In this work, the authors present a new Deep Learning, low-end device-friendly end-to-end approach for the classification of Normal patients, Covid positive, and pneumonia patients [14, 15]. All the networks trained on a public dataset with the same training and testing approach, the hyperparameters used for the models remained the same for all the models during the experiment for a fair comparison [16–19]. Results are evaluated by taking the average of five trials for all the models. Two approaches are used in our work.

i. The validation set and testing set are split in the 20% and 10% ratio, respectively. Data-Augmentations like rotation and flipping are applied to increase the data by a factor of 2.

ii. To focus on the abnormalities in the chest x-ray images caused by coronavirus, an end-to-end new deep convolutional neural network is proposed consisting of two attention mechanisms, namely spatial attention and channel attention.

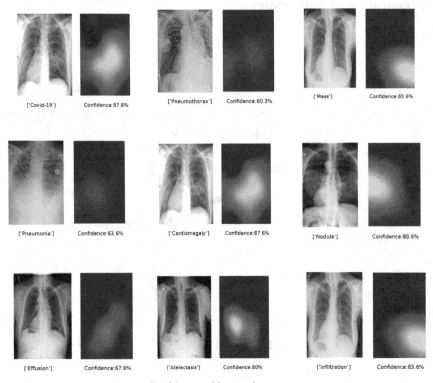

Confidence with attention maps

A New efficient module, Intermediate-Residual-SE-CBAM (IRSC), is proposed to improve the feature learning capability of the network by using a minimum number of parameters [19]. The models are tested on a new combined dataset called Covi-Xray. All the images of X-rays are taken from three open-source datasets, 1. Covid-19 Radiology Database 2. Covid-Chestxray-Dataset 3. ChexPert-Dataset A total of 7k images are created in the Covi-Xray dataset, which includes 3,000 covid-19 samples, 2,000 Viral pneumonia samples, 2000 Atelectasis, Cardiomegaly, Effusion, Infiltration, Mass, Nodule and Pneumothorax.. All the images are resized to 720 × 720, and three classes of Covid+, and Pneumonia are created. To account for the data imbalance, we apply data-augmentation procedures like vertical and horizontal flip during training procedure on Viral Pneumonia and Normal cases to increase their sample factor by 1.5 [20–22].

In this paper our main objective is to develop an efficient model which can run on FPGA's to perform validation on the chest x-ray images which can help in early diagnosis of lung related diseases. As implementation of AI for validation using FPGA's is a booming field, we try to develop a parameter efficient model as the hardware's can't support heavy models.

2   Preparation of Your Paper

(See Fig. 2).

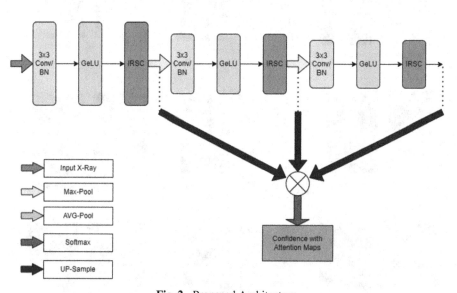

Fig. 2. Proposed Architecture

2.1   GeLU Activation Function

The Gaussian Error Linear Unit aims to combine the neuron outputs of the network. ReLU, ELU, and PReLU are fast and provide better convergence for the neural networks than the sigmoid functions [23, 24]. Dropout procedures and regularization mechanisms multiply these activations by zero. GeLU aims to combine both methods, and a new regularizer, namely Zone out, multiplies the input given by Eq. (1),

$$m \sim \text{Bernolli}(\varphi(x)) \tag{1}$$

where φ,

$$\varphi(x) = P(X < x) \text{and } N \sim (0, 1) \tag{2}$$

P is the probability distribution used; the reason behind choosing this distribution depends on the fact that the input follows a normal distribution after the Batch-Normalizing operation. To convert the output from stochastic to deterministic, we find the expected value of the transformation, $E(mx) = xE(x)$, where m is the Bernoulli random variable its expected value is (x). The function (x) is often referred to as a commutative distribution of the Gaussian function. Therefore, the GeLU function can be given in Eq. (3)

$$A = \text{GELU}(x) = xP(X \leq x) = x\varphi(x) \approx 0.5x(1 + \tanh[\sqrt{(2/\pi}} (x + 0.044715x^3)]) \tag{3}$$

Fig. 3. GeLU activation block

2.2 Intermediate Residual SE CBAM (IRSC) Block

The structural architecture of the IRSC block is shown in Fig. 3. It consists of naïve stacking of two convolutional layers with 3×3 filters accompanied by Batch-normalizing layers and ReLU activation layers. An intermediate SE layer follows up the output of these layers. It consists of a convolutional transformation filter Ctr; the filter is used for transformation from X to Y, as illustrated in Fig. 4. By taking Xo as the input to the SE block, we write the output of the transformation Ctr as $Y = [x1, x2,....xc]$. Where xc is given by Eq. (4).

$$x_c = k_c * X = \sum_{s=1}^{c'} k_c^s * x^s \tag{4}$$

Here, $K = [k1, k2,.... kc]$ denotes the filter kernels where kc denotes the parameters of the c-th filter, The term kcs Depicts a 2-D kernel in space representing each channel kc that acts on the X. To maintain simplicity, all the bias terms are ignored in the above equation. H, W represents the height and width of input Xo and H' and W' represents the height and width of image feature after applying transformation Ctr.

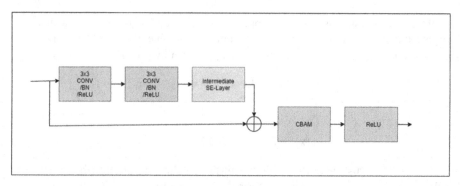

Fig. 4. Structure of IRSC block

The feature enhancement using the SE filter is done in two stages; the first stage, squeezing, involves using a statistic function called Z. A statistic ZRC is given by shrinking the transformed output through its spatial dimensions in height H and width W, where Zc is given as in Eq. (5),

$$Z_c = C_{sq}(x_c) = \frac{1}{H \times W} \sum\nolimits_{i=1}^{H} \sum\nolimits_{j=1}^{W} y_c(i, j) \qquad (5)$$

The second step involves using gating operation with a sigmoid function which is given by Eq. (6),

$$t = C_{ex}(Z, W) = \sigma(g(Y, W)) = \sigma(W_2 \delta(W_1 Y)) \qquad (6)$$

Here W1 R(C/r × C) and W2 R(C × C/r) and represents the ReLU function. As illustrated in Fig. 3, input Z performs channel description with global average pooling. There are 2 fully connected layers. The two FC layers are designed with dimensionality reduction ratio r using, the reduction ratio used throughout the network topology is 2, and is given by Eq. (7),

$$\frac{2}{r} \sum\nolimits_{s=1}^{S} N_s \cdot C_s^2 \qquad (7)$$

Here, Ms represents the number for stage s, and Cs is the number of output channels. The overall output of the SE block is given by Eq. (8),

$$\tilde{A} = C_{scale}(y_c, s_c) = s_c \cdot y_c \qquad (8)$$

The scaled feature from the SE block is added with the residual input to form the output of the residual intermediate SE block given by Eq. (9) and Fig. 5 shows the structure of squeeze and excitation module.

$$O = x_o + \tilde{A} \qquad (9)$$

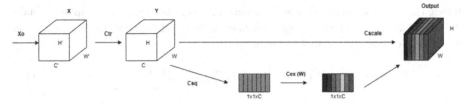

Fig. 5. Diagram of Squeeze and Excitation module.

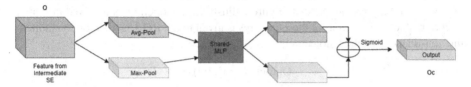

Fig. 6. Diagram of Channel Attention Mechanism.

2.3 Attention Mechanism

The output from the intermediate SE block is given to two separate attention blocks, namely channel and spatial attention. The attention mechanism uses the melded Cross-channel and spatial information to give more dominant and essential features from the images. This separate attention module allows the feature learning of the model to get sophisticated attention features using both the spatial and the channel weights of the feature scale from the SE block. The scaled map got from the Intermediate SE block is passed through the attention mechanism. The idea is to produce the channel attention feature maps by exploring the inter-channel relations from the scaled input (O) by considering each channel as a detector. For the practical computation of the channel, attention squeezing is performed on the input feature(O). For the aggregation of the spatial information, average-pooling and softmax operations are performed simultaneously to gather important clues of the distinctive features of channel-wise attention. These ideas are empirically confirmed that both softmax and average pooling significantly improve the network's representational power(cite CBAM paper). The channel attention module is shown in Fig. 6. Here Bc is represented as the output feature from the channel-attention block. The aggregated spatial information of the feature maps is obtained by using both average-pool and max-pooling activations, and this generates two distinct descriptors, namely RavgC and RmaxC, which denote the average pooled max-pooled features, respectively. Both of the descriptors are given to the shared network to give Oc as the channel attention map. Here OcRCx1x1, the shared network involves a multi-layer perceptron (MLP) with a single hidden layer. The output Oc is given by,

$$O_c = \sigma \left(MLP \left(R_{avg}^c \right) \right) + MLP \left(R_{max}^c \right) \tag{10}$$

Here, denotes the sigmoid operation; the weights are shared between the inputs and the ReLU activation functions. The spatial map is generated by using the differences in spatial relationships of the features. We apply average pooling and maxpooling operations through the axis and concatenate both the generate a standard feature descriptor to obtain the spatial attention features. During concatenation operation, convolution operation is performed to generate a spatial feature map Os,

$$O_s(O) = \sigma\left(F^{7x7}\left[\left(R^s_{avg}; R^s_{max}\right)\right]\right) \quad (11)$$

where is a sigmoid operator, F7 × 7 represents the convolution of the 7x7 filter. The 2D maps generated from aggregating the channels is given by, Ravgs Rmaxs. The former represents the average-pool feature across the channel, and the latter represents the max-pooled feature across the channel. Figure 7 illustrates the operations involved in spatial attention block. The output of these features is given to the ReLU activation to maintain the non-linearity of the whole attention system, given by,

$$\text{Output} = \alpha(O_s(O)) \quad (12)$$

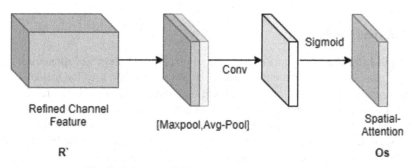

Fig. 7. Diagram of Channel Attention Mechanism..

3   Results and Discussions

This paper compares the quantitative and qualitative results of our proposed model with cutting-edge models like ResNet, DenseNet, Mobile-Net, MNAS-Net, and VGG-16. All the performance evaluations are done over five trials, from that, the best results were adopted.

3.1   Setup

To perform this experiment, we used an online-browser based platform called Google Colab Notebook. All the trails are run on Nvidia K80 Graphic Processing Unit(GPU). It has a memory capacity of 12 Gb and operates with 0.82 GHz of memory clock. It can compute up to 4.1 FLOPS with 2 CPU cores embedded inside. The X-ray images of three classes. Covid+, Normal, and Pneumonia, are passed through the network's input, maintaining a batch size of 32 throughout the experiment. The batch size is selected keeping in mind the execution time and the GPU memory available for the experiment.

All the codes are run on the PyTorch platform. The images are extracted by using a Class that inherits from torch.utils.data.Dataset. The utils.data.DataLoader module is used to load the batches for training. All the convolutional operators, including activations, are inherited from the "nn" module from the PyTorch library. The image size of the x-ray used throughout the experiment is 280×280; the size is chosen depending on the resources available and the resolution required for the experiment's success. Each trial consisted of 25 epochs; Adam optimizer with learning rate 3e−5 is used to update the weights of the network parameters. To calculate the cost function, a multi-class Cross-Entropy loss function is used. The built-in Cross-Entropy function on PyTorch from the torch.nn module applies a softmax activation function to normalize and sum all the inputs to 1 before calculating the overall for the particular epoch. The equation gives the loss function,

$$Loss = -\sum_{i=1}^{N} t_i \log_e (I(x)_i) \qquad (13)$$

Here, N is the batch-size (N=32), ti is the target label, and I(x) is the softmax function given by,

$$I(x) = \frac{e^{x_i}}{\sum_{j=1}^{N} e^{x_j}} \qquad (14)$$

3.2 Evaluation Methods

All the modern performance evaluation methods are used to determine the performance of each model. With Accuracy, parameters like precision, recall, and IOU are also important

3.2.1 Accuracy

It involves the overall ratio of the predictions that are falsely classified positive and falsely classified negative for the context of classifying between Covid-positive patients, Normal patients, and viral Pneumonia positive patients. The overall accuracy is given by ratio in Eq. (15) (Fig. 8),

$$Accuracy = \frac{Number\ of\ correct\ predictions}{Total\ number\ of\ predictions} = \frac{TP + TN}{TP + TN + FP + FN} \qquad (15)$$

3.2.2 Precision and Recall

To calculate precision and recall, type1 and type2 are be defined. Type1 errors are called False positives, i.e., rejection of all the true null hypotheses, and type2 are called False negatives, which involve not rejecting the false null hypothesis. Precision can be seen as the percentage of relevant results, and Recall is the results that the model correctly

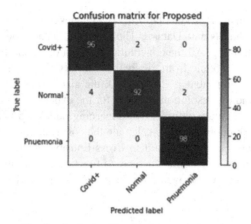

Fig. 8. Confusion Matrix of Proposed model

classifies. Therefore, the precision and recall can be given by the Eq. (16) and Eq. (17) respectively,

$$Precision = \frac{TruePositive}{ActualResults} = \frac{TP}{TP + FP} \tag{16}$$

$$Recall = \frac{True\ Positive}{Actual\ Results} = \frac{TP}{TP + FN} \tag{17}$$

3.2.3 F1-Score

F1-score evaluation parameter takes both precision and recalls into account to give a combined performance result. The score tries to allot more weights to false positives and false negatives and is given by Eq. (18),

$$F1 - score = 2 * \frac{Precision * Recall}{Precision + Recall} \tag{18}$$

3.2.4 IOU

IOU is considered one of the critical performance values for the classification task. Formula to calculate IOU is given by Eq. (19),

$$IOU = \frac{TP}{TP + FP + FN} \tag{19}$$

3.3 Classification Results

The performance and parameter comparisons are illustrated in Tables 1 and 2 respectively. Our proposed model uses fewer parameters and uses efficient modules like

Squeeze and Excitation to achieve state-of-the-art performance. Due to the over-fitting of data due to fewer data samples, over-fitting is common in deep-neural networks. All the layers of the stateof-the-art models for comparison are taken, except in the case of VGG-19, only one fully connected layer has been used. Comparing our proposed model with the VGG-19, our model achieves the highest increase of 13.14, 12.9, 15, 16, and 25.03 in Accuracy, Precision, Recall, F1, and IOU, respectively, the VGG-19 architecture. Our model uses 43 times lesser parameters than VGG-19. On a one-to-one comparison with ResNet50, our model uses 50 times lesser parameters and achieves an approximately 4-percent increase over Accuracy, Precision, Recall, and F1-score and a 6.96-percent increase in the Jaccard Index. DenseNet is a bench-mark model for the experiment which uses considerably fewer parameters with a high accuracy factor. Our proposed model uses 15.33 times lesser parameters and a performance increase of approximately 2-percent on all the parameters, including a 3.3-percent increase in IOU. Mobile-NetV2 and MNAS-Net are two models specifically designed for use in end devices like mobiles. These two models use the least amount of parameters than other state-of-art models with good performance in classification tasks. Our model uses close to 2M lesser parameters than both the models and achieves a 6% increase in Accuracy, Precision, Recall, and F1-Score and 10-percent in the IOU parameter. Figures 9, 10, 11, 12, 13 and 14 show the loss vs accuracy graph of Proposed, Resnet-18, DenseNet-121, MobileNet-V2, MNAS-Net and VGG-19 respectively.

Table 1. Comparison Table

Model	Accuracy	Precision	Recall	F1-Score	IOU
VGG-19	83	83.1	81	80	67.53
ResNet-18	92	92	92	92	85.6
DenseNet-121	94.33	94.33	94.33	94.33	89.26
MobileNet-V2	90.29	90.33	90.33	91	81.82
MNAS-Net	90.5	91	90	91	82.83
Proposed	96.14	96	96	96	92.5

Table 2. Parameter Table

Model	Parameters
VGG-19	20 M
ResNet-18	23 M
DenseNet-121	6.9 M
MobileNet-V2	3.5 M
MNAS-Net	3.1 M
Proposed	0.45 M

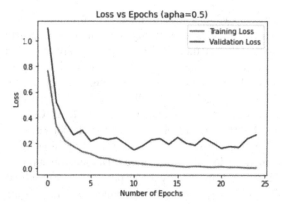

Fig. 9. Loss Vs Epoch of Proposed.

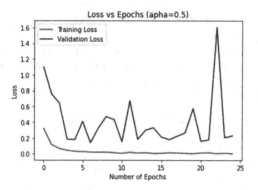

Fig. 10. Loss Vs Epoch of ResNet.

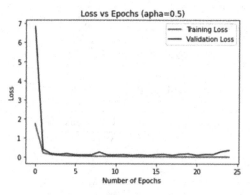

Fig. 11. Loss Vs Epoch of DenseNet.

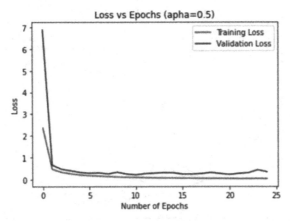

Fig. 12. Loss Vs Epoch of MobileNet.

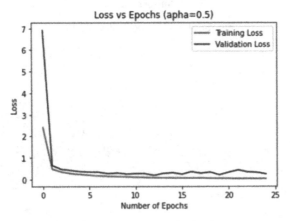

Fig. 13. Loss Vs Epoch of MNAS Net.

3.4 Analysis using DcGAN

DCGAN has some architectural constraints, and exhibited that they are a state of the art candidate when it comes to unsupervised learning [25]. Visual quality of samples generated from the generative image models has increased a lot, with concerns of memorizing and overfitting in the medical data of training samples rising. To show how our proposed model scales with data and high level and low level resolution generation, analysis is done using DcGAN. We use DcGAN to get extra features from the results got from Attention-CoviNet, we calculate error rate, accuracy and accuracy per class for each of the model. The features extracted from the models are given as input to the DcGAN network. The resolution of the image used as input is 240 × 240, the features are scaled to give the maximum feature value from the model. The Table 3, shows that our proposed model has the least error rate compared to other models. Table 4, shows the accuracy and accuracy per 100 images for all the models (Fig. 15).

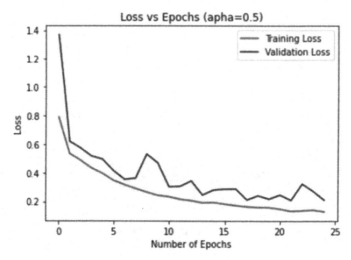

Fig. 14. Loss Vs Epoch of VGG-19.

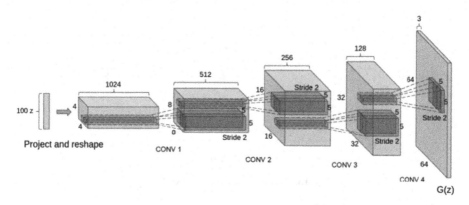

Fig. 15. DCGAN generator used for analysis [25].

Table 3. DCGAN Error rate analysis

CLASSES	VGG-19	RESNET-18	Dense-121	Mobile-Net V2	MNAS-Net	Proposed
Atelectasis	44.8% (±0.7%)	28.6% (±0.7%)	24.8% (±0.7%)	25.9% (±0.7%)	25.5% (±0.7%)	25.5% (±0.7%)
Cardiomegaly	45.5% (±0.7%)	26.2% (±0.7%)	25.1% (±0.7%)	25.4% (±0.7%)	25.3% (±0.7%)	23.7% (±0.7%)

(continued)

Table 3. (*continued*)

CLASSES	VGG-19	RESNET-18	Dense-121	Mobile-Net V2	MNAS-Net	Proposed
Covid-19	31.9% (±0.7%)	22.7% (±0.7%)	21.9% (±0.7%)	20.3% (±0.7%)	20.1% (±0.7%)	20.8% (±0.7%)
Effusion	65.7% (±0.7%)	38.6% (±0.7%)	35.3% (±0.7%)	36.9% (±0.7%)	36.7% (±0.7%)	30.3% (±0.7%)
Infiltration	43.9% (±0.7%)	24.7% (±0.7%)	23.7% (±0.7%)	22.5% (±0.7%)	22.1% (±0.7%)	21.8% (±0.7%)
Mass	47.3% (±0.7%)	28.7% (±0.7%)	27.7% (±0.7%)	27.7% (±0.7%)	26.9% (±0.7%)	23.7% (±0.7%)
Nodule	50.6% (±0.7%)	30.3% (±0.7%)	30.1% (±0.7%)	29.9% (±0.7%)	29.2% (±0.7%)	28.4% (±0.7%)
Pneumothorax	61.1% (±0.7%)	42.4% (±0.7%)	41.1% (±0.7%)	40.3% (±0.7%)	39.8% (±0.7%)	38.8% (±0.7%)
Pnuemonia	73.4% (±0.7%)	46.7% (±0.7%)	43.4% (±0.7%)	41.4% (±0.7%)	41.2% (±0.7%)	40.9% (±0.7%)

Table 4. DCGAN Accuracy Analysis

Model	Accuracy	Accuracy(100 per class)
VGG-19	84.10%	62.6%(0.7% t)
ResNet-18	92.50%	80.3%(0.7% t)
DenseNet-121	94.32%	84.7%(0.7% t)
MobileNet-V2	90.29%	82.4%(0.7% t)
MNAS-Net	90.60%	82.6%(0.7% t)
Proposed	96.25%	87.2%(0.7% t)

4 Conculsion

For better performance of Deep Learning architectures, there is a requirement of datasets involving more number of samples. Due to the high cost and unavailability of the dataset, we have to depend on procedures like data augmentation and regularization to avoid the networks from over-fitting the data. This paper proves that due to fewer data samples for training the neural networks, over-fitting occurs. To avoid the problem by incorporating a network with cost-effective modules. Due to the fewer parameters, the model's ability to learn decreases, so feature-enhancing modules like Squeeze and Excitation and attention mechanisms are used. Our model achieved the best Accuracy of 96.14%, best performance in Precision, Recall, and F1-score by obtaining 96%, and obtaining 92.56 in IOU. In the future, we focus our research work on CT images of the chest for the classification of Covid-19. The computational complexity of the architecture is reduced by using 452,361 parameters in total, thereby decreasing the training and testing time for the model.

References

1. Lecun, Y., Bottou, L., Bengio, Y., Haffner, P.: Gradient-based learning applied to document recognition. Proc. IEEE **86**(11), 2278–2324 (1998). https://doi.org/10.1109/5.726791
2. Shervin, M., Rahele, K., Milan, S., Shakib, Y., Ghazaleh, J.S.: Deep-COVID: predicting COVID-19 from chest X-ray images using deep transfer learning. Med. Image Anal. **65**, 101794 (2020). https://doi.org/10.1016/j.media.2020.101794
3. Karhan, Z., Akal, F.: Covid-19 classification using deep learning in chest X-ray images. Med. Technol. Congress **2020**, 1–4 (2020). https://doi.org/10.1109/TIPTEKNO50054.2020.9299315
4. Jabber, B., Lingampalli, J., Basha, C.Z., Krishna, A.: Detection of Covid-19 patients using chest x-ray images with convolution neural network and mobile net. In: 2020 3rd International Conference on Intelligent Sustainable Systems (ICISS), pp. 1032–1035 (2020). https://doi.org/10.1109/ICISS49785.2020.9316100
5. Hernandez, D., Pereira, R., Georgevia, P.: COVID-19 detection through X-Ray chest images. In: 2020 International Conference Automatics and Informatics (ICAI), pp. 1–5 (2020). https://doi.org/10.1109/ICAI50593.2020.9311372
6. Tabik, S., et al.: COVIDGR dataset and COVID-SDNet methodology for predicting COVID-19 based on chest X-Ray images. IEEE J. Biomed. Health Inform. **24**(12), 3595–3605 (2020). https://doi.org/10.1109/JBHI.2020.3037127
7. Huang, G., Liu, Z., Van Der Maaten, L., Weinberger, K.Q.: Densely connected convolutional networks. In: 2017 IEEE Conference on Computer Vision and Pattern Recognition (CVPR), pp. 2261–2269 (2017). https://doi.org/10.1109/CVPR.2017.243
8. He, K., Zhang, X., Ren, S., Sun, J.: Deep residual learning for image recognition. In: 2016 IEEE Conference on Computer Vision and Pattern Recognition (CVPR), pp. 770–778 (2016). https://doi.org/10.1109/CVPR.2016.90
9. Szegedy, C., et al.: Going deeper with convolutions. In: 2015 IEEE Conference on Computer Vision and Pattern Recognition (CVPR), pp. 1–9 (2015). https://doi.org/10.1109/CVPR.2015.7298594
10. Zhong, X., Gong, O., Huang, W., Li, L., Xia, H.: Squeeze-and-excitation wide residual networks in image classification. In: 2019 IEEE International Conference on Image Processing (ICIP), pp. 395–399 (2019). https://doi.org/10.1109/ICIP.2019.8803000
11. Li, C., et al.: Attention Unet++: a nested attention-aware U-Net for liver CT image segmentation. In: 2020 IEEE International Conference on Image Processing (ICIP), pp. 345–349 (2020). https://doi.org/10.1109/ICIP40778.2020.9190761
12. Zhu, Y., Zhao, C., Guo, H., Wang, J., Zhao, X., Lu, H.: Attention CoupleNet: fully convolutional attention coupling network for object detection. IEEE Trans. Image Process. **28**(1), 113–126 (2019). https://doi.org/10.1109/TIP.2018.2865280
13. Zhang, Q., et al.: Dense attention fluid network for salient object detection in optical remote sensing images. IEEE Trans. Image Process. **30**, 1305–1317 (2021). https://doi.org/10.1109/TIP.2020.3042084
14. Hu, J., Shen, L., Sun, G.: Squeeze-and-excitation networks. In: 2018 IEEE/CVF Conference on Computer Vision and Pattern Recognition, pp. 7132–7141 (2018). https://doi.org/10.1109/CVPR.2018.00745

15. Thange, U., Shukla, V.K., Punhani, R., Grobbelaar, W.: Analyzing COVID-19 dataset through data mining tool "Orange". In: 2021 2nd International Conference on Computation, Automation and Knowledge Management (ICCAKM), pp. 198–203 (2021). https://doi.org/10.1109/ICCAKM50778.2021.9357754

16. Umri, B.K., Wafa Akhyari, M., Kusrini, K.: Detection of Covid-19 in Chest X-ray Image using CLAHE and Convolutional Neural Network. In: 2020 2nd International Conference on Cybernetics and Intelligent System (ICORIS), pp. 1–5 (2020). https://doi.org/10.1109/ICORIS50180.2020.9320806

17. Tan, M., et al.: MnasNet: platform-aware neural architecture search for mobile. In: 2019 IEEE/CVF Conference on Computer Vision and Pattern Recognition (CVPR), pp. 2815–2823 (2019). https://doi.org/10.1109/CVPR.2019.00293

18. Sinha, D., El-Sharkawy, M.: Thin MobileNet: an enhanced mobilenet architecture. In: 2019 IEEE 10th Annual Ubiquitous Computing, Electronics & Mobile Communication Conference (UEMCON), pp. 0280–0285 (2019). https://doi.org/10.1109/UEMCON47517.2019.8993089

19. Liu, S., Deng, W.: Very deep convolutional neural network based image classification using small training sample size. In: 2015 3rd IAPR Asian Conference on Pattern Recognition (ACPR), pp. 730–734 (2015). https://doi.org/10.1109/ACPR.2015.7486599

20. Chowdhury, M.E.H., et al.: Can AI help in screening viral and COVID-19 pneumonia? IEEE Access 8, 132665–132676 (2020). https://doi.org/10.1109/ACCESS.2020.3010287

21. Rahman, T., et al.: Exploring the effect of image enhancement techniques on COVID-19 detection using chest X-ray images. Comput. Biol. Med. 132, 104319 (2021). ISSN 0010-4825, https://doi.org/10.1016/j.compbiomed.2021.104319

22. Jeremy, I., et al.: CheXpert: a large chest radiograph dataset with uncertainty labels and expert comparison. In: Proceedings of the AAAI Conference on Artificial Intelligence, vol. 33, pp. 590–597. https://doi.org/10.1609/aaai.v33i01.3301590

23. Surendran, R., Osamah, K., Carlos, A., Carlos, R.: Deep learning based intelligent industrial fault diagnosis model. Comput. Mater. Continua 70 (2021). https://doi.org/10.32604/cmc.2022.021716

24. Ding, B., Qian, H., Zhou, J.: Activation functions and their characteristics in deep neural networks. In: 2018 Chinese Control And Decision Conference (CCDC), pp. 1836–1841 (2018). https://doi.org/10.1109/CCDC.2018.8407425

25. Alec, R., Luke, M., Soumith, C.: Unsupervised Representation Learning with Deep Convolutional Generative Adversarial Networks (2015)

A Deep Learning Framework for Violence Detection in Videos Using Transfer Learning

Gurmeet Kaur$^{(\boxtimes)}$ and Sarbjeet Singh

Department of Computer Science and Engineering, Panjab University,
Chandigarh, India
grmtkaur76@gmail.com

Abstract. Violence detection in videos has become a significant problem in the field of computer vision. It involves the process of automatically identifying violent behavior in video content. The rapid growth of digital media led the researchers to focus on developing effective methods for detecting violence that can automatically identify real-world instances of violence in order to maintain public safety and security. This paper presents a transfer learning approach for detection of violence in videos. The approach uses a pre-trained ResNet50, a deep residual network to extract the features from frames of videos. The results show that the suggested approach achieved accuracy of 98.89% on Hockey Fight and 99.97% on Movies datasets and highlighting the effectiveness of transfer learning in learning discriminatory features for recognition of violent action in videos over traditional hand-crafted feature detectors.

Keywords: Violence Detection · Computer Vision · Transfer Learning · ResNet50

1 Introduction

Detection of violence in video is a very important task for maintaining world security and stability. The abundance of video data generated every second has made it unrealistic to manually monitor and capture every violent scene in real time. With the advancements in technology and rise in digital media, the need for automatic violence detection systems has become increasingly important [14]. These systems can help in real-time monitoring of surveillance videos and alert authorities in case of any violent incidents. Hence, violence detection has become emerging field of computer vision and machine learning that focuses on automatically identifying instances of violent behavior in video content. The field is rapidly evolving due to its applications in monitoring public spaces, analyzing security camera footage, monitoring social media content, and more. Researchers have developed various approaches for recognizing violent/aggressive movements by learning visual patterns and benchmark datasets describing violence in videos.

P. Das et al. (Eds.): AMRIT 2023, CCIS 1954, pp. 100–111, 2024.
https://doi.org/10.1007/978-3-031-47221-3_9

Earlier approaches for violence detection mostly relied on hand-defined feature descriptors. In these approaches, feature representations such as Motion SIFT (MoSIFT), Interest Points, Motion blobs and features, blood and flame detection are typically used for violence identification and to differentiate between fight scenes and normal ones [3,17,20]. In recent times, deep learning has played a vital role in the evolution of computer vision, which has made significant achievements in a number of fields including object detection, image recognition, and facial recognition, human activity recognition and so on. Similarly, deep learning techniques like CNN, 3DCNN, LSTM and ConvoLSTM have gained prominence in the field of violence detection by providing significant improvements over conventional feature-based techniques related to accuracy and robustness [6,15,22–24].

However, developing deep learning models from scratch require huge computational power and large amounts of problem-specific data. Development of a large annotated dataset is labor-intensive and time-consuming, which is a considerable obstruction for developing deep learning models. In order to overcome these limitations, researchers have introduced the concept of transfer learning within deep learning. An effective transfer learning strategy involves fine-tuning a CNN model pre-trained on a particular dataset for a new task, even in another domain, by reducing the requirement of large volume of labeled data and computational resources, making the training process more efficient [16]. With the evolution of research in this area, more and more researchers have applied transfer learning to decrease the time and effort needed to develop new violence detection methods [1,5,26]. Based on this, we present a an model for detecting violence in video sequences using ResNet50 [11]. The model obtained the accuracy of98.89% on Hockey Fight [3] and 99.97% on Movies [3] datasets. The key objectives of this paper are as:

- To propose a transfer learning-based approach for violence detection.
- To extract deep spatial features from video sequences and fine-tune the model with these features to classify violent and non-violent items.
- To compare the performance of the proposed approach with existing methods in terms of learning fine-grained features, robustness, and accuracy.

In this work, a transfer learning-based violence detection approach is proposed. Deep spatial features are extracted from the frames of violent and non-violent video sequences using ResNet50 [11] and model is fine tuned on Hockey dataset [3], movies dataset [3] to learn these features for classification of violent and non violent scenes. The result of using deep representation model shows the comparative performance to the existing approaches, in terms of learning fine-grained features, robustness and accuracy.

The remainder of the paper is structured as: Sect. 2 presents an overview of previous work and approaches in the field of violence detection. Section 3 outlines the proposed methodology in detail, followed with Sect. 4 discussing the experiments and analyses carried out to evaluate the model's performance. Finally, Sect. 5 summarizes the work presented in this work and states the conclusions and future work in violence detection research.

2   Related Work

Over the years, researchers have proposed a variety of techniques to detect violent behavior in video content. In early studies in this domain, scientists typically used visual cues, such as blood or flames, as well as audio cues, like screams, gunshots and breaking sounds [4,9,28]. Unfortunately, these audio contents are often not present in surveillance videos, making it necessary to focus on video content alone. As a result, video-based approaches have become more widely used in recent times, and they can be categorized into two categories based on the feature extraction method used: traditional based on hand-crafted features and based on deep learning.

In traditional approaches, researchers manually define a set of features that are relevant for detecting violence, such as motion patterns, object interactions, and facial expressions. The features are then extracted from the video data and used as inputs to a machine learning model. Based on STIP and MoSIFT descriptors, a novel framework for fight detection was developed by Bermejo et al. [3]. Another study [18] analysed ocular data using LaSIFT descriptors and Lagrangian methods to obtain movement features. With the help of optical flow, a descriptor named ViF was developed for the detection of real-time violence in crowd using a new dataset of crowded scenes [10]. Unfortunately, this method performs poorly while dealing with non-crowding scenes. Gao et al. [8] proposed an extended ViF descriptor (OViF). ViF and OViF are both effective at classifying data in crowded environments, but ViF offers greater classification ability than to OViF alone. However, combination of both even greater accuracy than either algorithm on its own. [29] Proposed to extract two descriptors called LHOG and LHOF. The author suggested to identify motion areas and calculate both descriptors on these areas and then coded using Bag of Words (BoW) model. In [7] presented a method based on movement filtering and motion boundary SIFT for detecting violence.

As deep learning has become increasingly popular in the field of visual action recognition, so researchers also developed models that detect violent acts in videos using deep neural networks. Sultani et al. [24] developed ranking framework by utilizing 3D convolution features for detection of normal and anomalous data. Three-dimensional convolutional neural network is used for extraction of features from videos, then classified them as positive or negative bags based on ranking. In [27], a multi-feature fusion approach is proposed, in which static features are extracted from frames by pseudo-3D (P3D) convolution networks and temporal features by Long-Short term memory. Singh et al. [21] constructed a DNN-based model that utilized a CNN for extraction of spatial features from individual frames, followed by a Convolutional Long-Short Term Memory (ConvLSTM) to predict future motions. Similarly, Randon et al. developed a space-time encoder framework named ViolenceNet for the detecting violent actions by integrating a customized DenseNet with a multi-head self-attention and a BiConvLSTM modules. Some other researchers [1,13,25] used pre-trained CNN to develop models for violent human behaviors detection. In [25], authors proposed a noble technique by exploring three pre-trained CNNs- VGG16, VGG19, and ResNet50. This technique utilizes thirty frames drawn from videos to extract features and feeds these features into a LSTM network along with attention layers.

Based on the above discussed approaches, we leverages the concept of transfer learning and present a architecture for detection of violence in videos with ResNet50 a pre-tarined CNN.

3   Proposed Approach

The purpose of this study is to propose a deep learning model for detection of violent events in video clips. An overview of proposed approach is shown in Fig. 1. Initially, the input videos are processed by pre-processing operations. Then at next stage a deep network extract and learn spatial features for each frame. These extracted features are then fed into fully connected layers for classification. Detailed procedures for our approach are outlined in subsections below.

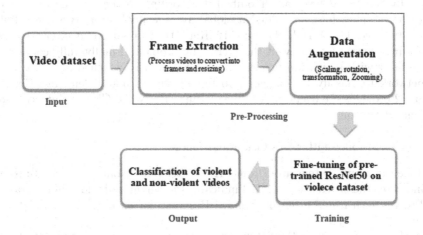

Fig. 1. Overview of proposed approach for violence detection

3.1   Preprocessing

During this step, all input videos go through a series of operations and conversions, before it is processed by ResNet. As a first step, frame extraction is performed on the videos, and each video's frame count is ensured to be a multiple of five in order to prevent duplicate frames. Then every frame for input videos is resized to fixed size of 224×224×3. Afterward, some techniques for data augmentation such as rotation, horizontal flips, zooming, horizontal and vertical shifts are applied to image data which helps to prevent over fitting and enhance the applicability of the model. To simplify the process of extracting meaningful features from the data, data scaling and normalization are also performed. Finally, labelled data is shuffled and divided into 75% for training and 25% as a testing set.

3.2   Fine-Tuning of ResNet50

In this work, we utilized ResNet50 a deep residual neural network as pre-trained backbone with features learned from ImageNet dataset of 1 million images and 1000 different classes. ResNet50 is 50 layers deep network composed of residual connections with 25 million parameters, which allow it to bypass the problem of vanishing gradients and to learn more complex representations of the input data by speed up the training time [11]. We evaluated multiple architectures for detecting violent content in videos, including ResNet18, ResNet50, and ResNet101. But ultimately decided to use ResNet50 due to its favorable balance between performance and computational efficiency. By considering the above advantages, we applied transfer learning to retrain the ResNet50 CNN by fine-tuning its hyperparameters and parameters for violence detection. There are 26,211,714 parameters in the model, of which 26,158,594 are trainable and 53,120 are not trainable. For this task, we freezed the initial layers of the network and used it as base model to extract the features. Subsequently, we removed the last fully connected layer and modified it according to the selected dataset of two classes. Therefore, we added a global average pool layer, two dense fully connected layer (1024, ReLU[1]) and (512, ReLU). Then a dropout layer attached to the base network to prevent the overfitting problem. Finally, fully connected dense layer with activation function named Sigmoid is attached as a last layer of network to classify the images into two classes as non-violent and violent. Figure 2 provides the framework of the modified ResNet50 CNN architecture for violence detection.

The steps followed for this task are as follow:

1. Take the input 3D tensor of (frame, H, W, RGB) as (None, 244, 244, 3); None indicates the batch size, 224*224 indicates the height and width of the frame.
2. Process each frame using pre-trained Resnet50 for extraction of spatial features.
3. Fifth stage (conv5) of ResNet50 provides a feature map of 3D tensor of size (None, 7, 7, 2048); 7 and 7 the spatial dimensions of the feature map, and 2048 representing the number of channels.
4. Apply the global average on this output to get a 1d tensor with same number of channels in last step.
5. Pass the output of the previous step into the sequence fully connected layer with activation Relu to apply a non-linear transformation to the input features to enhance their representational power
6. Finlay the output of the previous step feed into the output layer i.e fully connected dense layer with sigmoid activation to classify the video based on the probability of violence in it.

---

[1] Rectified Linear Unit

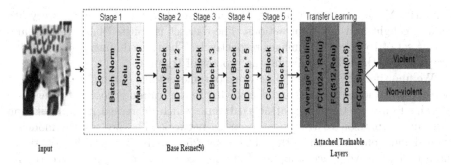

Fig. 2. The modified ResNet50 architecture for violence detection

4   Experiments and Results

We utilized Python 3.7 and the TensorFlow 2.0 package along with other libraries to accomplish this task. The implementation is performed on Google Colaboratory to access GPU runtime and Google Drive for storing and accessing the dataset. The model is evaluated using two benchmark dataset i.e Hockey Fight [3] and Movie [3] datasets.

4.1   Datasets

1. *Hockey Fight:* The dataset has 1000 videos captured from NHL games with real-life violent incidents, with 500 of them violent and 500 non-violent. The dataset is considered to be a challenging dataset for violence detection due to the complex and dynamic nature of the events, as well as the large number of variables such as camera angles, lighting conditions, and player movements.
2. *Movies:* The dataset refers to a collection of 200 clips, split into 100 non-violent and 100 violent categories. The violent videos were gathered from movies, where as non-violent clips are collected from normal activities. The average resolution of the Movie dataset is 360×250 pixels.

4.2   Experiment Settings

The ResNet model employed to extract spatial information from the images. Therefore, video datasets are converted to their respective frames in order to train the model effectively. Hockey dataset consist of 9003 labelled images for both activities(fight and non-fight), from which 6752 are selected for training of model and 2251 as a validation set to evaluate the model's performance.

Similarly, Movies dataset contains 2004 annotated images by saving every 5th frame for fight and non-fight actions. The dataset is splitted into training and testing set of 1503 and 501 images respectively. The resized images of 244*244*3 are passed to the network during the process of training.

We fine-tuned the ResNet50 network parameters like batch size 34, learning rate of 0.001, momentum 0.9 and decay 0.01. In the training process, the network is trained only on the newly attached layers, which decreased its computation costs in comparison to training the entire network. The pre-trained model is trained independently on both datasets for 25 epochs, but it will stop training if no loss convergence is found after three consecutive epochs. Furthermore, Stochastic gradient descent (SGD), called Adam, is applied as a otimizer during training to minimize the crossentropy loss.

4.3 Result

In this study, the attached layers of ResNet50 are trained end-to-end for detecting violence and classify videos into violent and non-violent categories for 25 iterations and evaluated the model with accuracy as metric. The optimized model achieved accuracy 98.89% for hockey dataset after 21 epochs and 99.7% on movie dataset after 18 epochs. Training and validation accuracy and loss graphs of this model for both datasets are shown in Fig. 3.

Moreover, the model's performance is evaluated using the metrics shown as classification report in Table 1 and Table 2 for Hockey and movies datasets individually. Figure 4 shows the quantitative results for each dataset in terms of confusion matrices. There are 55 misclassified samples on the hockey dataset out of 2251 samples, and 23 on the movies dataset out of 501 samples. For hockey dataset, the number of misclassified samples is higher than those of non-violence, whereas model correctly predicts every sample of violence for movie dataset.

Table 1. Classification report for Hockey fight dataset

Class	Precision	Recall	F1-Score	Support
NonViolence	0.96	0.99	0.98	1125
Violence	0.99	0.96	0.98	1126

Table 2. Classification report for Movie dataset

Class	Precision	Recall	F1-Score	Support
NonViolence	1.00	0.91	0.95	252
Violence	0.92	1.00	0.96	249

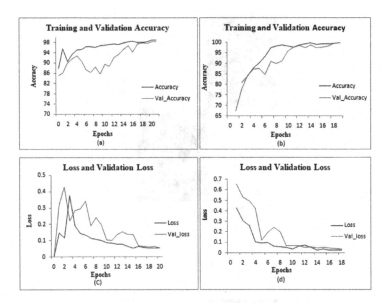

Fig. 3. The performance of ResNet-50 architectures on Hockey and Movies datasets: (a) training and validation accuracy for Hockey fight (b) training and validation accuracy for Movies(c) training and validation loss for Hockey fight (d) training and validation loss for Movies

4.4   Comparison and Discussion

This subsection evaluates the accuracy of the proposed method in comparison to state-of-the-art approaches based on these datasets. In 2017, Kecceli et al. [12] developed a method utilizing optical flow and pre-trained CNN. They achieved 94.40% accuracy on hockey dataset while we achieved 98.8% on same. An approach combining 2D-CNN with Hough Forest features was proposed by Serrano et al. [19]. They evaluated their approach elevated on Hockey and Movies datasets and got accuracies 94.6±0.6%, 99±0.5% correspondingly. Another article employed a pre-trained VGG-16 for extracting spatial features, followed by an LSTM for extraction of temporal features, which achieved 88.2% as benchmark accuracy with the introduction of a new dataset named Real- Life Violence Situations [22]. In [13], the author proposes a Fine-tuned MobileNet framework for detecting violence in videos, but experiments indicate it is less accurate than that proposed. Similarly, Asad et al. [2] proposed the method for learning motion and appearance features by merging the feature maps with convolutional and recurrent networks. Table 3 presents the accuracy comparison of existing state-of-the-arts with proposed approach.

Using ResNet50 deep model with pre-trained imagenet weights, the proposed approach provide significant performance on both datasets. Results show the

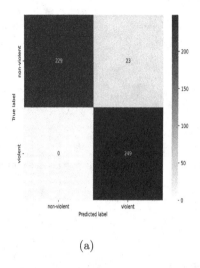

(a)

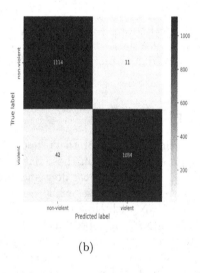

(b)

Fig. 4. Confusion Matrices (a) Movies (b) Hockey Fight

highest accuracy of 98.87% on hockey dataset and 99.97% on movies dataset. In addition to performing on movie dataset with homogeneous content, the model can also learn deep features for constant motion among players and dynamic backgrounds in hockey fight dataset. In our study, we did not evaluate the model on any crowd violence dataset, so we do not know how it performs in complex social situations like crowds.

Table 3. Accuracy comparison of existing methods with proposed approach

Year	Method	Hockey Fight	Movies
2017	Kecceli et al. [12]	94.4%	96.5%
2018	Ismael et al. [19]	94.6%	99.0%
2019	Soliman et al. [22]	95.1%	99.0%
2019	Khan et al. [13]	87.5%	99.5%
2021	Asad et al. [2]	98.8%	99.1%
–	**Proposed**	**98.8%**	**99.9%**

5   Conclusion

In this research, a deep learning model is suggested to solve the problem of violence recognition by utilizing the idea of transfer learning. For the purpose of identifying violence in videos, a CNN network ResNet50 is used for extracting spatial features, while the last fully connected layer is replaced with additional trainable layers. Finally, fine-tuned the model on two video datasets - Hockey and Movies. The results of the study showed that transfer learning with the ResNet50 model is effective in detecting violence in video clips with comparable performance to current ate-of-the-art models. Results of this study demonstrate the potential of transfer learning in violence detection by eliminating the issue of over-fitting in training a deep network from scratch.

In future work, the current approach can be extended to other domains such as crowd violence, violent events in social media and other real-life scenarios. The proposed model was tested on only two datasets in the current study, so future research could examine a approach on other real-life and more diverse datasets. Furthermore, adding new layers, including convolutional and recurrent layers, can be considered to enhance model performance.

References

1. Abdali, A.M.R., Al-Tuma, R.F.: Robust real-time violence detection in video using CNN and LSTM. In: 2019 2nd Scientific Conference of Computer Sciences (SCCS), pp. 104–108. IEEE (2019). https://doi.org/10.1109/SCCS.2019.8852616
2. Asad, M., Yang, J., He, J., Shamsolmoali, P., He, X.: Multi-frame feature-fusion-based model for violence detection. Vis. Comput. **37**, 1415–1431 (2021). https://doi.org/10.1007/s00371-020-01878-6
3. Bermejo Nievas, E., Deniz Suarez, O., Bueno García, G., Sukthankar, R.: Violence detection in video using computer vision techniques. In: Real, P., Diaz-Pernil, D., Molina-Abril, H., Berciano, A., Kropatsch, W. (eds.) CAIP 2011. LNCS, vol. 6855, pp. 332–339. Springer, Heidelberg (2011). https://doi.org/10.1007/978-3-642-23678-5_39
4. Cheng, W.H., Chu, W.T., Wu, J.L.: Semantic context detection based on hierarchical audio models. In: Proceedings of the 5th ACM SIGMM International Workshop on Multimedia Information Retrieval, pp. 109–115 (2003). https://doi.org/10.1145/973264.973282

5. Choudhary, R., Solanki, A.: Violence detection in videos using transfer learning and LSTM. In: Verma, P., Charan, C., Fernando, X., Ganesan, S. (eds.) Advances in Data Computing, Communication and Security. LNDECT, vol. 106, pp. 51–62. Springer, Singapore (2022). https://doi.org/10.1007/978-981-16-8403-6_5

6. Ding, C., Fan, S., Zhu, M., Feng, W., Jia, B.: Violence detection in video by using 3D convolutional neural networks. In: Bebis, G., et al. (eds.) ISVC 2014. LNCS, vol. 8888, pp. 551–558. Springer, Cham (2014). https://doi.org/10.1007/978-3-319-14364-4_53

7. Febin, I., Jayasree, K., Joy, P.T.: Violence detection in videos for an intelligent surveillance system using MoBSIFT and movement filtering algorithm. Pattern Anal. Appl. **23**(2), 611–623 (2020). https://doi.org/10.1007/s10044-019-00821-3

8. Gao, Y., Liu, H., Sun, X., Wang, C., Liu, Y.: Violence detection using oriented violent flows. Image Vis. Comput. **48**, 37–41 (2016). https://doi.org/10.1016/j.imavis.2016.01.006

9. Giannakopoulos, T., Makris, A., Kosmopoulos, D., Perantonis, S., Theodoridis, S.: Audio-visual fusion for detecting violent scenes in videos. In: Konstantopoulos, S., Perantonis, S., Karkaletsis, V., Spyropoulos, C.D., Vouros, G. (eds.) SETN 2010. LNCS (LNAI), vol. 6040, pp. 91–100. Springer, Heidelberg (2010). https://doi.org/10.1007/978-3-642-12842-4_13

10. Hassner, T., Itcher, Y., Kliper-Gross, O.: Violent flows: real-time detection of violent crowd behavior. In: 2012 IEEE Computer Society Conference on Computer Vision and Pattern Rrecognition Workshops, pp. 1–6. IEEE (2012). https://doi.org/10.1109/CVPRW.2012.6239348

11. He, K., Zhang, X., Ren, S., Sun, J.: Deep residual learning for image recognition. In: Proceedings of the IEEE Conference on Computer Vision and Pattern Recognition, pp. 770–778 (2016). https://doi.org/10.48550/arXiv.1512.03385

12. Keçeli, A., Kaya, A.: Violent activity detection with transfer learning method. Electron. Lett. **53**(15), 1047–1048 (2017). https://doi.org/10.1049/el.2017.0970

13. Khan, S.U., Haq, I.U., Rho, S., Baik, S.W., Lee, M.Y.: Cover the violence: a novel deep-learning-based approach towards violence-detection in movies. Appl. Sci. **9**(22), 4963 (2019). https://doi.org/10.3390/app9224963

14. Li, J., Jiang, X., Sun, T., Xu, K.: Efficient violence detection using 3D convolutional neural networks. In: 2019 16th IEEE International Conference on Advanced Video and Signal Based Surveillance (AVSS), pp. 1–8. IEEE (2019). https://doi.org/10.1109/AVSS.2019.8909883

15. Mohtavipour, S.M., Saeidi, M., Arabsorkhi, A.: A multi-stream CNN for deep violence detection in video sequences using handcrafted features. Vis. Comput., 1–16 (2022). https://doi.org/10.1007/s00371-021-02266-4

16. Mumtaz, A., Sargano, A.B., Habib, Z.: Violence detection in surveillance videos with deep network using transfer learning. In: 2018 2nd European Conference on Electrical Engineering and Computer Science (EECS), pp. 558–563. IEEE (2018). https://doi.org/10.1109/EECS.2018.00109

17. Nam, J., Alghoniemy, M., Tewfik, A.H.: Audio-visual content-based violent scene characterization. In: Proceedings 1998 International Conference on Image Processing. ICIP98 (Cat. No. 98CB36269). vol. 1, pp. 353–357. IEEE (1998). https://doi.org/10.1109/ICIP.1998.723496

18. Senst, T., Eiselein, V., Sikora, T.: A local feature based on lagrangian measures for violent video classification. In: 6th International Conference on Imaging for Crime Prevention and Detection (ICDP-15), pp. 1–6. IET (2015). https://doi.org/10.1049/ic.2015.0104

19. Serrano, I., Deniz, O., Espinosa-Aranda, J.L., Bueno, G.: Fight recognition in video using hough forests and 2D convolutional neural network. IEEE Trans. Image Process. **27**(10), 4787–4797 (2018). https://doi.org/10.1109/TIP.2018.2845742

20. Serrano Gracia, I., Deniz Suarez, O., Bueno Garcia, G., Kim, T.K.: Fast fight detection. PLoS ONE **10**(4), e0120448 (2015). https://doi.org/10.1371/journal.pone.0120448

21. Singh, P., Pankajakshan, V.: A deep learning based technique for anomaly detection in surveillance videos. In: 2018 Twenty Fourth National Conference on Communications (NCC), pp. 1–6. IEEE (2018). https://doi.org/10.1109/NCC.2018.8599969

22. Soliman, M.M., Kamal, M.H., Nashed, M.A.E.M., Mostafa, Y.M., Chawky, B.S., Khattab, D.: Violence recognition from videos using deep learning techniques. In: 2019 Ninth International Conference on Intelligent Computing and Information Systems (ICICIS), pp. 80–85. IEEE (2019). https://doi.org/10.1109/ICICIS46948.2019.9014714

23. Sudhakaran, S., Lanz, O.: Learning to detect violent videos using convolutional long short-term memory. In: 2017 14th IEEE International Conference on Advanced Video and Signal based Surveillance (AVSS), pp. 1–6. IEEE (2017). https://doi.org/10.48550/arXiv.1709.06531

24. Sultani, W., Chen, C., Shah, M.: Real-world anomaly detection in surveillance videos. In: Proceedings of the IEEE Conference on Computer Vision and Pattern Recognition, pp. 6479–6488 (2018). https://doi.org/10.48550/arXiv.1801.04264

25. Sumon, S.A., Goni, R., Hashem, N.B., Shahria, T., Rahman, R.M.: Violence detection by pretrained modules with different deep learning approaches. Vietnam J. Comput. Sci. **7**(01), 19–40 (2020). https://doi.org/10.1142/S2196888820500013

26. Sumon, S.A., Shahria, M.D.T., Goni, M.D.R., Hasan, N., Almarufuzzaman, A.M., Rahman, R.M.: Violent crowd flow detection using deep learning. In: Nguyen, N.T., Gaol, F.L., Hong, T.-P., Trawiński, B. (eds.) ACIIDS 2019. LNCS (LNAI), vol. 11431, pp. 613–625. Springer, Cham (2019). https://doi.org/10.1007/978-3-030-14799-0_53

27. Xu, X., Wu, X., Wang, G., Wang, H.: Violent video classification based on spatial-temporal cues using deep learning. In: 2018 11th International Symposium on Computational Intelligence and Design (ISCID), vol. 1, pp. 319–322. IEEE (2018). https://doi.org/10.1109/ISCID.2018.00079

28. Zajdel, W., Krijnders, J.D., Andringa, T., Gavrila, D.M.: CASSANDRA: audio-video sensor fusion for aggression detection. In: 2007 IEEE Conference on Advanced Video and Signal based Surveillance, pp. 200–205. IEEE (2007). https://doi.org/10.1109/AVSS.2007.4425310

29. Zhou, P., Ding, Q., Luo, H., Hou, X.: Violence detection in surveillance video using low-level features. PLoS ONE **13**(10), e0203668 (2018). https://doi.org/10.1371/journal.pone.0203668

Multi-focus Image Fusion Methods: A Review

Ravpreet Kaur$^{(\boxtimes)}$ and Sarbjeet Singh

UIET, Panjab University, Chandigarh 160014, India
`ravpreet3@pu.ac.in, sarbjeet@pu.ac.in`

Abstract. Because of the limitations of cameras optical lenses, which have finite Depth-of-Field (DOF), acquiring images that are completely in focus is a challenging task. In particular, the scene contents inside the scope of the DOF are focused whereas those outside the scope are out of focus. Therefore to create a fully focused image, a number of algorithms have been developed by researchers with the aim of fusing together a number of partly focused images of the same view. These algorithms are generally categorized into conventional and deep learning-based techniques. Multi-focus image fusion (MFIF) has several applications such as optical microscopy, micro-image fusion, digital photography and remote sensing networks. In this review, first, the concept of image fusion has been explained followed by the categories of MFIF including the limitations of traditional image fusion methods. Subsequently, the experimental evaluation of seven MFIF methods is performed both qualitatively and quantitatively on Real-MFF dataset. As a final conclusion, the paper discusses some challenges and areas for future research.

Keywords: Depth-of-field · Multi-focus image fusion · Conventional · Deep learning · Real-MFF dataset

1 Introduction

The purpose of image fusion is to integrate significant data from either two source images or more of an identical view into one image that is more reliable than each of the input images and better appropriate for both human and machine vision [1,15].

In image processing applications, image capturing is a crucial step. The digital single-lens reflex camera cannot grab every detail of the scene; hence, certain details could unavoidably be lost. Usually, cameras can focus upto a finite distance. This is owing to the restricted DOF of optical lenses of conventional cameras because the cameras have the ability to concentrate only on a particular region in an image while keeping the remainder of the scene out of focus [5,40].

Image objects captured in the range of limited DOF appear to be clear or focused, while those far from the focus point seem blurry or out-of-focus [34].

P. Das et al. (Eds.): AMRIT 2023, CCIS 1954, pp. 112–125, 2024.
https://doi.org/10.1007/978-3-031-47221-3_10

The quality of an image can be enhanced by using a method that allows for the collection of images where each item is focused. This is accomplished by blending numerous images taken from various focal planes into one image. A fused image with every item in focus is what is produced. This process is known as Multi-focus image fusion (MFIF) as demonstrated in Fig. 1. MFIF is a way of extending the DOF by combining partly focused images of a similar view into a full focused image. It is a challenging problem, while being intriguing and appearing to be simple [34,37].

As a result of human perception, clear images are more informative than the blurred ones [45]. The three different levels can be used to carry out the fusion process listed as follows [5,15,41].

Source image A   Source Image B   Fused Image

Fig. 1. The MFIF procedure. Image courtesy of [42]

Pixel level - The pixels in the input images are used directly in this type of image fusion. The pixels of source images are averaged or maximized to produce the fused image which is less computationally intensive and highly accurate. It is the lowest level of image fusion [5,15].

Feature level - It is also called fusion at intermediate level. Firstly, it extracts the features like textures, corners, edges etc. from the source images and then use it for further processing and analysis to perform fusion [15].

Decision level - It is the highest level of image fusion where input images are separately processed to extract the information. The decision-level has finest real-time performance when compared to pixel and feature-level, but information loss is one of the major downside of this image fusion level [41].

The remaining sections of this paper is divided as follows: Sect. 2 describes the methods of MFIF both traditional and deep learning based. Section 3 describes the dataset used for evaluation and examines the performance of different MFIF methods. Conclusion along with future work is summarized in last section.

2   Multi-focus Image Fusion Methods

As the demand for fully focused images has increased, many MFIF algorithms have been developed to enhance the image quality. In general, these algorithms are classified into two kinds: Traditional and deep learning-based MFIF methods. These two categories are explained as below [9,45]:

2.1 Traditional MFIF Methods

Several MFIF approaches are broadly divided into transform domain and spatial domain-based methods in accordance with the fusion strategy as shown in Fig. 2.

Spatial Domain Based Methods

In these methods, the input images are blended using spatial features of the images. The focus level or sharpness of an image is precisely estimated using pixel intensity values. As opposed to transform domain-based approaches, the inverse transform stage is not used in spatial domain. A spatial domain based approach consists of establishing the weight maps for each input images and then to use the weighted average of those images to construct a fused image. However, there are significant drawbacks, such as spatial distortion, noise sensitivity, and artefacts that show up in the fused image. There are three types of spatial domain-based methods which are explained as below [5, 24, 31]:

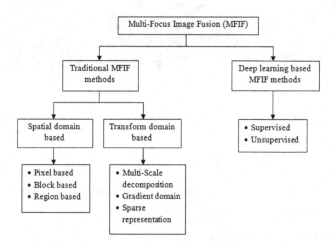

Fig. 2. Classification of MFIF methods [24, 41, 43]

a. Pixel-based methods

Pixel-based MFIF techniques have drawn a significant interest lately. These techniques may often retain the spatial integrity of the fused image while retrieving sufficient detail from the original images [31]. The most basic pixel-based image fusion technique simply averages each source image's pixel values. Even though it is quick and easy, the fused image tends to have some blurring effects [9]. Several techniques have been put forth for fusing images using pixel-based methods like [3, 6, 23, 28, 33].

b. Block-based methods

In block-based approaches, each pair of input images is partitioned into a set of blocks of specific dimension. The ALM[1] is constructed for each block and

---

[1] Activity Level Measurement.

then every block is fused in accordance with it. The result of the fusion technique is significantly influenced by the block size because these approaches suffer from blocking effects due to the concurrent existence of focused and defocused blocks [5,31]. Some of these block-based methods are [4,12,20,32] etc. Many researchers have suggested various improved block-division algorithms to address the issue of block size such as [2,30,44].

c. Region-based methods

Based on the concept of image segmentation, region-based techniques put a great deal of emphasis on the accuracy of segmentation. While the architecture of these methods is analogous to that of block-based methods, the fundamental distinction between them is that the former executes ALM in every segmented area of non-uniform size [24]. Like block-based approaches, these methods also suffer from blocking effects. Some representative methods include [8,13,19].

Transform-Domain Based Methods

These methods differ from the methods based on spatial domain; in that they convert the input images into a different domain where they can be correctly fused. As shown in Fig. 3, these methods are composed of three phases. First, an image decomposition method is used to convert the input image into transform coefficients. Second, the image fusion activity is carried out in three steps: ALM, fusion rule and CV^2. Third, the fused image is rebuilt by applying inverse transform. The more the layers of decomposition, more detailed the information, but the efficiency will degrade. As a result, perfectly managing the relation between the decomposition layer and execution efficiency, fusion effect will be significantly increased [24,46]. The categories of transform domain-based methods include MSD^3, SR^4 and GD^5-based methods.

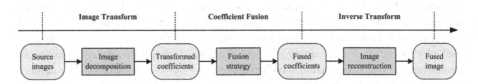

Fig. 3. Simplified representation of transform domain-based methods [24]

a. MSD-based methods

MSD-based techniques are currently often employed in the domain of MFIF. The three forms of MSD have been identified: pyramid transform, multi-scale geometric analysis and wavelet transform. Due to their superior feature

---

[2] Consistency Verification.

[3] Multi-scale decomposition.

[4] Sparse representation.

[5] Gradient domain.

representation capabilities, these approaches have shown tremendous success in the MFIF field [5]. The MSD approach has the benefit of obtaining more precise details of feature which leads to better fusion impact overall but there is excessive breakdown of information resulting in a significant amount of computation [24,46].

b. SR-based methods

In SR-based methods, the signals are depicted by a small number of atoms extracted from a predetermined lexicon. In the domain of MFIF, sparse representation methods have become a prominent area. The problem of noise in fused image can be better resolved by sparse representation. Furthermore, the method is complex, has limited computing efficiency and performs poorly in real-time [5,46].

c. GD-based methods

Fusing the gradient depiction of the input images is the fundamental aim of gradient domain-based image fusion techniques. Then fused image is recreated by using the fused gradient. This method can enhance the visual effect of image by preserving the features [24].

Limitations of Traditional MFIF Methods

It is widely known that ALM (also known as focus measure) and fusion rule are two important elements for both transform and spatial domain based image fusion techniques. But these conventional methods have some problems in handling them.

Firstly, feature extraction from various source images must be performed using the same transformation to ensure that successive feature fusion is viable. Since the extracted features do not consider the distinctive differences between source images, they may not be as informative as they could be. Secondly, the majority of these methods frequently use far more intricate fusion rules, which may not be appropriate for use in real-world scenarios. Thirdly, The manual design of both focus measure and fusion rule is a challenging task that requires a lot of experience. Furthermore, the conventional techniques separate the focus measurement and fusion rule, which restricts the performance of fusion, as creating an optimal design that takes into account all the required parameters is nearly impractical [16,18,21,39,45].

However it does not imply that these intricately developed fusion rules and ALM are not significant contributions. The issue is that handcrafted layout is actually rather difficult. Additionally, from a particular perspective, it is hard to create a perfect design that takes into consideration all the required components [21].

To address the above limitations, recently deep learning based approaches have received much attention. They have made a great advances in the image fusion field. These methods can take advantage of different network branches to extract more accurate features. They learn the focus measure by using the deep characteristics that are more potent than the handcrafted ones. Additionally, during training, these methods learn fusion rules automatically. Furthermore, they can manage ALM and fusion rules collectively through learning, making it more capable than traditional MFIF methods. As a result, deep learning-based methods lead to improved fusion performance in MFIF problem. Under the direction of well-designed loss functions, deep learning-based approaches may develop a more logical feature fusion strategy to attain adaptive feature fusion. Deep learning accelerates the development of image fusion by taking use of these benefits and outperforming conventional approaches in terms of performance [17, 41, 43].

2.2 Deep Learning (DL)-Based Methods

In the realm of MFIF, deep learning-based approaches have emerged as a very prominent domain and lately achieved cutting-edge results by utilising neural networks' potent image representation capabilities. Since 2017, a plethora of deep learning-based MFIF techniques have been presented, resulting in a new trend in this area. Deep learning techniques are further classified into supervised and unsupervised deep learning based techniques [24, 43].

a. Supervised DL-based MFIF methods
   Supervised methods needs the labelled data to train the network and to generate the labeled data, some methods uses Gaussian blur strategy to blur the clear image while others employ matte boundary defocus mode to give better fusion results by providing realistic training data [43]. Some representative supervised methods include [9, 12, 18, 21, 27, 39, 45].
b. Unsupervised DL-based MFIF methods
   Recently, ConvNet (CNNs) have been employed for MFIF. Since there isn't enough labelled data for supervised learning, so existing techniques uses Gaussian blur to imitate defocus and create synthetic training data with ground-truth. Further, they categorize the pixels as clear or blur and use the resulting fusion weight maps for post-processing. Thus, unsupervised MFIF methods are employed [37]. Some of the unsupervised methods are [11, 14, 26, 35, 37].

Readers are advised to go through reference [5] and [43] for in-depth coverage of MFIF methods.

3 Experimental Findings

This section presents a comparative analysis of the multi-focus image fusion algorithms. Four objective metrics were used to determine the performance of 7 familiar image fusion methods on 5 multi-focus image pairs. All the algorithms were run using the default settings listed by their authors.

3.1 Dataset and MFIF Methods for Comparison

In this study, 5 multi-focus image pairs are randomly selected from Real-MFF dataset [42] for evaluation as demonstrated in Fig. 4. The dataset consists of two

source image pairs of size 433×625 along with ground-truth. In addition to this, 7 prevalent MFIF techniques are employed for comparison out of which 4 are based on traditional approaches namely ASR [25], DSIFT [23], CSR [22], GFDF [29] and 3 are based on deep learning viz. CNN [21], MADCNN [16], DRPL [18]
.

(a)   (b)   (c)   (d)   (e)

Fig. 4. Five sets of source images from Real-MFF dataset used in our experiment.

3.2   Qualitative and Quantitative Evaluation

Quantitative Evaluation

To determine the efficacy of fused images obtained using different MFIF algorithms, the objective evaluation metrics were used. There are four types of metrics: metrics based on (i) information theory, (ii) image structural similarity, (iii) image feature and (iv) human perception. Four metrics were examined in this paper, each from a different category [24,43]. These are described as below:

- *Mutual information* (Q_{MI}) [10] is an information theory based metric that assesses the quantity of information carried over from input images to the fused image.
- *The Gradient based metric* $(Q_{AB/F})$ [36] determines to which degree the edge details from source images is incorporated into the fused image. It is metric based an image feature.
- *Yang's metric* (Q_Y) [38] which determines how much structural detail is retained in the fused image, is an image structural similarity-based metric.
- *Human perception* (Q_{CB}) [7] measures the degree to which key features of the human vision are similar.

For all four assessment metrics listed above, greater the value, greater is the quality of image fusion. The implementation of these metrics is contained in the MATLAB toolbox[6] provided by the first author of [26]. Table 1 shows the

---

[6] https://github.com/zhengliu6699/imageFusionMetrics.

objective performance of several image fusion techniques. Each metric's top two best results are shown in bold.

DRPL and DSIFT perform effectively in retaining information in a fused image. Although the results of both the above techniques are very close to each other but DSIFT lacks in retaining the features from source images as can be seen from Q_{CB} metric. GFDF and CNN performs better when compared to other methods. The performance of MADCNN is average. ASR and CSR does not perform well enough in comparison to other methods.

Table 1. The objective comparison of the state-of-the-art image fusion methods. The average values is shown across all 5 pairs of Real-MFF dataset.

Methods	Q_{MI}	$Q_{AB/F}$	Q_Y	Q_{CB}
ASR	1.0786	0.7351	0.9634	0.8369
DSIFT	**1.1984**	0.7542	0.9785	**0.8640**
CSR	0.5760	0.5046	0.8407	0.5961
GFDF	1.1725	**0.7552**	**0.9800**	**0.8646**
CNN	1.1637	**0.7547**	**0.9792**	0.8631
MADCNN	1.1549	0.7536	0.9771	0.8563
DRPL	**1.1984**	0.7540	0.9781	0.8609

Qualitative Evaluation

For qualitative evaluation, results of only one multi-focus image pair is presented in Fig. 5 due to page limitation. This figure includes source image pairs as well as the results of different state-of-the-art image fusion techniques on Real-MFF dataset [42]. Here the selected area is marked with red boxes and the Fig. 6 displays the zoomed-in selected regions. For better comparison, difference of images is calculated by subtracting source image B displayed in Fig. 5(b) from each fused image. The difference images presented in Fig. 7. has been adjusted to the range of 0 to 1.

It is clearly seen from Fig. 7(a), that ASR represents some minute residuals in the background between the focused part of input image B and the fused image. The branches of the tree are faintly visible in the background of difference image. It also suffers from artifacts around the edges of leaves. The CSR technique does not perform well enough as it introduces sharpness around the edges of leaves along with the increase in the intensity of the fused image (see Fig. 5(d) and 6(d)). The background is clearly visible which results in the overall low image fusion quality as depicted in Fig. 7(b).

The DSIFT method in Fig. 7(c), process the boundaries poorly near the clear and blur region of the fused image (see the magnified Fig. 5(e) and 6(e)). The corner edge of the leaves are blurred and somewhere they are not complete as compared to other methods. GFDF in Fig. 7 (d) performs much better than DSIFT but fails near the boundary of leaves due to the existence of artifacts

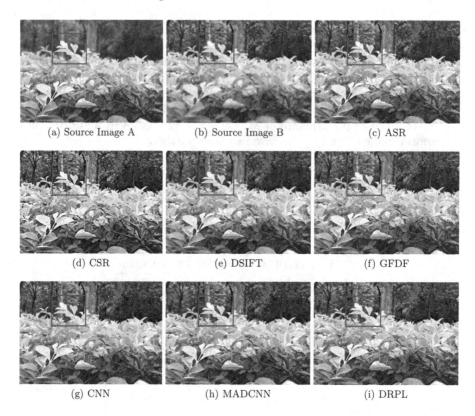

Fig. 5. Qualitative image fusion outcomes of different fusion algorithms on one MFIF pair from Real-MFF dataset

and some minor residuals. The visual comparison of CNN in Fig. 6(g) and 7(e) is almost similar to GFDF as both the techniques have some slight residuals around the top leaf. MADCNN in Fig. 7(f) also has good image fusion quality except that some part of the edges are blurred and few artifacts are present on the edge boundary of leaves. In Fig. 7(g), the DRPL outperforms all other methods as edge boundary of leaves can be precisely seen resulting in a superior image fusion quality (see Fig. 5(i) and 6(i)).

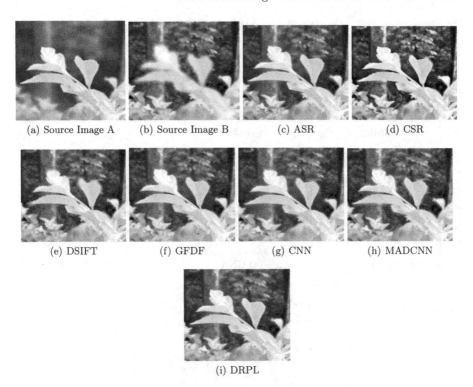

(a) Source Image A   (b) Source Image B   (c) ASR   (d) CSR

(e) DSIFT   (f) GFDF   (g) CNN   (h) MADCNN

(i) DRPL

Fig. 6. Magnified sections of images shown in Fig. 5(a) to 5(i)

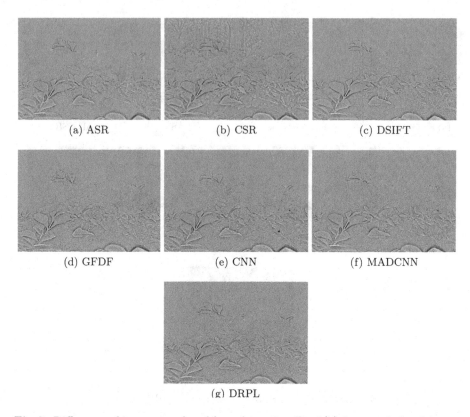

(a) ASR   (b) CSR   (c) DSIFT

(d) GFDF   (e) CNN   (f) MADCNN

(g) DRPL

Fig. 7. Difference of images produced by subtracting Fig. 5(b) from each fused image (Fig. 5(c)-5(i))

4   Conclusion and Future Directions

In this review, both traditional and DL-based MFIF methods have been discussed along with their characteristics. Experiments were conducted to estimate the performance of 7 MFIF algorithms on 5 pairs selected at random from the Real-MFF dataset. Four evaluation metrics have been used. The results demonstrate that DL-based MFIF methods perform well when compared to traditional MFIF methods.

Although MFIF has been the subject of significant research over the past few years, there are still some challenges that needs to be addressed such as reducing the level of noise, reducing the running time of the process, developing a real-time image fusion algorithm which leads to broader application prospects, designing effective fusion schemes for fusing boundary regions.

In the future, the aforementioned problems can be considered for building an efficient MFIF approach. A researcher can consider combining more than two multi-focus images. Furthermore, it will be extremely beneficial if an MFIF approach can be built for certain application cases.

References

1. Anish, A., Jebaseeli, T.J.: A survey on multi-focus image fusion methods. Int. J. Adv. Res. Comput. Eng. Technol. (IJARCET) **1**(8), 2012 (2012)
2. Aslantas, V., Kurban, R.: Fusion of multi-focus images using differential evolution algorithm. Expert Syst. Appl. **37**(12), 8861–8870 (2010). https://doi.org/10.1016/j.eswa.2010.06.011
3. Bai, X., Liu, M., Chen, Z., Wang, P., Zhang, Y.: Multi-focus image fusion through gradient-based decision map construction and mathematical morphology. IEEE Access **4**, 4749–4760 (2016). https://doi.org/10.1109/ACCESS.2016.2604480
4. Bai, X., Zhang, Y., Zhou, F., Xue, B.: Quadtree-based multi-focus image fusion using a weighted focus-measure. Inf. Fusion **22**, 105–118 (2015). https://doi.org/10.1016/j.inffus.2014.05.003
5. Bhat, S., Koundal, D.: Multi-focus image fusion techniques: a survey. Artif. Intell. Rev. **54**(8), 5735–5787 (2021). https://doi.org/10.1007/s10462-021-09961-7
6. Bouzos, O., Andreadis, I., Mitianoudis, N.: Conditional random field model for robust multi-focus image fusion. IEEE Trans. Image Process. **28**(11), 5636–5648 (2019). https://doi.org/10.1109/TIP.2019.2922097
7. Chen, Y., Blum, R.S.: A new automated quality assessment algorithm for image fusion. Image Vis. Comput. **27**(10), 1421–1432 (2009). https://doi.org/10.1016/j.imavis.2007.12.002
8. Duan, J., Chen, L., Chen, C.P.: Multifocus image fusion with enhanced linear spectral clustering and fast depth map estimation. Neurocomputing **318**, 43–54 (2018). https://doi.org/10.1016/j.neucom.2018.08.024
9. Guo, X., Nie, R., Cao, J., Zhou, D., Qian, W.: Fully convolutional network-based multifocus image fusion. Neural Comput. **30**(7), 1775–1800 (2018). 0.1162/neco_a_01098
10. Hossny, M., Nahavandi, S., Creighton, D.: Comments on 'information measure for performance of image fusion'. Electron. Lett. **44**(18), 1066–1067 (2008). https://doi.org/10.1049/el:20081754
11. Hu, X., Jiang, J., Liu, X., Ma, J.: Zero-shot multi-focus image fusion. In: 2021 IEEE International Conference on Multimedia and Expo (ICME), pp. 1–6. IEEE (2021). https://doi.org/10.1109/ICME51207.2021.9428413
12. Huang, W., Jing, Z.: Multi-focus image fusion using pulse coupled neural network. Pattern Recogn. Lett. **28**(9), 1123–1132 (2007). https://doi.org/10.1016/j.patrec.2007.01.013
13. Huang, Y., Li, W., Gao, M., Liu, Z.: Algebraic multi-grid based multi-focus image fusion using watershed algorithm. IEEE Access **6**, 47082–47091 (2018). https://doi.org/10.1109/ACCESS.2018.2866867
14. Jung, H., Kim, Y., Jang, H., Ha, N., Sohn, K.: Unsupervised deep image fusion with structure tensor representations. IEEE Trans. Image Process. **29**, 3845–3858 (2020). https://doi.org/10.1109/TIP.2020.2966075
15. Kaur, H., Koundal, D., Kadyan, V.: Image fusion techniques: a survey. Arch. Comput. Methods Eng. **28**(7), 4425–4447 (2021). https://doi.org/10.1007/s11831-021-09540-7
16. Lai, R., Li, Y., Guan, J., Xiong, A.: Multi-scale visual attention deep convolutional neural network for multi-focus image fusion. IEEE Access **7**, 114385–114399 (2019). https://doi.org/10.1109/ACCESS.2019.2935006
17. Li, J., Yuan, G., Fan, H.: Multifocus image fusion using wavelet-domain-based deep cnn. Computational intelligence and neuroscience 2019 (2019). https://doi.org/10.1155/2019/4179397

18. Li, J., Guo, X., Lu, G., Zhang, B., Xu, Y., Wu, F., Zhang, D.: Drpl: deep regression pair learning for multi-focus image fusion. IEEE Trans. Image Process. **29**, 4816–4831 (2020). https://doi.org/10.1109/TIP.2020.2976190

19. Li, M., Cai, W., Tan, Z.: A region-based multi-sensor image fusion scheme using pulse-coupled neural network. Pattern Recogn. Lett. **27**(16), 1948–1956 (2006). https://doi.org/10.1016/j.patrec.2006.05.004

20. Li, S., Kwok, J.T., Wang, Y.: Combination of images with diverse focuses using the spatial frequency. Inf. Fusion **2**(3), 169–176 (2001). https://doi.org/10.1016/S1566-2535(01)00038-0

21. Liu, Y., Chen, X., Peng, H., Wang, Z.: Multi-focus image fusion with a deep convolutional neural network. Inf. Fusion **36**, 191–207 (2017). https://doi.org/10.1016/j.inffus.2016.12.001

22. Liu, Y., Chen, X., Ward, R.K., Wang, Z.J.: Image fusion with convolutional sparse representation. IEEE Signal Process. Lett. **23**(12), 1882–1886 (2016). https://doi.org/10.1109/LSP.2016.2618776

23. Liu, Y., Liu, S., Wang, Z.: Multi-focus image fusion with dense sift. Inf. Fusion **23**, 139–155 (2015). https://doi.org/10.1016/j.inffus.2014.05.004

24. Liu, Y., Wang, L., Cheng, J., Li, C., Chen, X.: Multi-focus image fusion: a survey of the state of the art. Inf. Fusion **64**, 71–91 (2020). https://doi.org/10.1016/j.inffus.2020.06.013

25. Liu, Y., Wang, Z.: Simultaneous image fusion and denoising with adaptive sparse representation. IET Image Proc. **9**(5), 347–357 (2015). https://doi.org/10.1049/iet-ipr.2014.0311

26. Ma, B., Zhu, Y., Yin, X., Ban, X., Huang, H., Mukeshimana, M.: Sesf-fuse: an unsupervised deep model for multi-focus image fusion. Neural Comput. Appl. **33**(11), 5793–5804 (2021). https://doi.org/10.1007/s00521-020-05358-9

27. Ma, H., Zhang, J., Liu, S., Liao, Q.: Boundary aware multi-focus image fusion using deep neural network. In: 2019 IEEE International Conference on Multimedia and Expo (ICME), pp. 1150–1155. IEEE (2019)

28. Nejati, M., Samavi, S., Shirani, S.: Multi-focus image fusion using dictionary-based sparse representation. Inf. Fusion **25**, 72–84 (2015). https://doi.org/10.1016/j.inffus.2014.10.004

29. Qiu, X., Li, M., Zhang, L., Yuan, X.: Guided filter-based multi-focus image fusion through focus region detection. Signal Process. Image Commun. **72**, 35–46 (2019). https://doi.org/10.1016/j.image.2018.12.004

30. Siddiqui, A.B., Jaffar, M.A., Hussain, A., Mirza, A.M.: Block-based pixel level multi-focus image fusion using particle swarm optimization. Int. J. Innov. Comput. Inf. Control **7**(7), 3583–3596 (2011)

31. Tang, H., Xiao, B., Li, W., Wang, G.: Pixel convolutional neural network for multi-focus image fusion. Inf. Sci. **433**, 125–141 (2018). https://doi.org/10.1016/j.ins.2017.12.043

32. Vakaimalar, E., Mala, K., et al.: Multifocus image fusion scheme based on discrete cosine transform and spatial frequency. Multimed. Tools Appl. **78**(13), 17573–17587 (2019). https://doi.org/10.1007/s11042-018-7124-9

33. Xia, X., Yao, Y., Yin, L., Wu, S., Li, H., Yang, Z.: Multi-focus image fusion based on probability filtering and region correction. Signal Process. **153**, 71–82 (2018). https://doi.org/10.1016/j.sigpro.2018.07.004

34. Xu, H., Fan, F., Zhang, H., Le, Z., Huang, J.: A deep model for multi-focus image fusion based on gradients and connected regions. IEEE Access **8**, 26316–26327 (2020). https://doi.org/10.1109/ACCESS.2020.2971137

35. Xu, H., Ma, J., Jiang, J., Guo, X., Ling, H.: U2fusion: A unified unsupervised image fusion network. IEEE Trans. Pattern Anal. Mach. Intell. **44**(1), 502–518 (2020). https://doi.org/10.1109/TPAMI.2020.3012548

36. Xydeas, C.S., Petrovic, V., et al.: Objective image fusion performance measure. Electron. Lett. **36**(4), 308–309 (2000)

37. Yan, X., Gilani, S.Z., Qin, H., Mian, A.: Unsupervised deep multi-focus image fusion. arXiv preprint arXiv:1806.07272 (2018)

38. Yang, C., Zhang, J.Q., Wang, X.R., Liu, X.: A novel similarity based quality metric for image fusion. Information Fusion **9**(2), 156–160 (2008). https://doi.org/10.1109/ICALIP.2008.4589989

39. Yang, Y., Nie, Z., Huang, S., Lin, P., Wu, J.: Multilevel features convolutional neural network for multifocus image fusion. IEEE Trans. Comput. Imaging **5**(2), 262–273 (2019). https://doi.org/10.1109/TCI.2018.2889959

40. Zafar, R., Farid, M.S., Khan, M.H.: Multi-focus image fusion: algorithms, evaluation, and a library. Journal of Imaging **6**(7), 60 (2020). https://doi.org/10.3390/jimaging6070060

41. Zhang, H., Xu, H., Tian, X., Jiang, J., Ma, J.: Image fusion meets deep learning: a survey and perspective. Inf. Fusion **76**, 323–336 (2021). https://doi.org/10.1016/j.inffus.2021.06.008

42. Zhang, J., Liao, Q., Liu, S., Ma, H., Yang, W., Xue, J.H.: Real-mff: a large realistic multi-focus image dataset with ground truth. Pattern Recogn. Lett. **138**, 370–377 (2020). https://doi.org/10.1016/j.patrec.2020.08.002

43. Zhang, X.: Deep learning-based multi-focus image fusion: a survey and a comparative study. IEEE Trans. Pattern Anal. Mach. Intell. (2021). https://doi.org/10.1109/TPAMI.2021.3078906

44. Zhang, X., Han, J., Liu, P.: Restoration and fusion optimization scheme of multi-focus image using genetic search strategies. Optica Applicata 35(4) (2005)

45. Zhao, W., Wang, D., Lu, H.: Multi-focus image fusion with a natural enhancement via a joint multi-level deeply supervised convolutional neural network. IEEE Trans. Circuits Syst. Video Technol. **29**(4), 1102–1115 (2018). https://doi.org/10.1109/TCSVT.2018.2821177

46. Zhou, Y., Yu, L., Zhi, C., Huang, C., Wang, S., Zhu, M., Ke, Z., Gao, Z., Zhang, Y., Fu, S.: A survey of multi-focus image fusion methods. Appl. Sci. **12**(12), 6281 (2022)

Cache Memory and On-Chip Cache Architecture: A Survey

Nurulla Mansur Barbhuiya$^{(\boxtimes)}$, Purnendu Das, and Bishwa Ranjan Roy

Department of Computer Science, Assam University, Silchar, Assam, India
nurullabarbhuiya@gmail.com

Abstract. Presently, one of the most essential performance of new multicore CPUs is processing speed. Various components, including cache, are employed to increase the processing speed of the processor. When cache memory comes to multi-core system speed, it is crucial. Because CPU speed is increasing at an alarming rate, an extremely fast cache memory is required to keep up with the processor. On-chip cache systems Cache memory is used to store information. Between the main memory and the CPU, there is a buffer called cache. The rate at which information flows between the central processor and main memory is synchronized using cache memory. The advantage of storing knowledge in cache over RAM is that it has faster retrieval times, but it has the downside of consuming on-chip energy. The performance of cache memory is evaluated in this research using these three variables: miss rate, miss penalty, and cache time interval.

Keywords: On-chip cache memory · addresses · hit ratio · miss ratio · cache memory

1 Introduction

The definition of cache is "a collection of like items kept in an extremely hidden or inaccessible spot." Caches, which are typically constructed of RAM, are at the top of the memory hierarchy RAM. The primary difference between a cache and other memory hierarchy levels is that a cache uses hardware to search memory locations, whereas other memories use software, or software and hardware combined, to do so. To improve the performance of the processors, raise their operating frequency or speed as well as the amount of work they complete per cycle. Several techniques for enhancing parallelism have arisen as a result of the growth in transistor count on a device. Microprocessors will soon be able to simultaneously run many processes or threads and take use of commonalities at the process or thread level.

In 1960, the IBM system/360 Model 85 was the first to use cache memory. An on-chip 8 KB L1 cache was initially introduced in the 1980s with the Intel 486DX microprocessor. Cache memories temporarily store the primary memory sections' contents that are most likely to be needed, which are compact, quick memories. Cache memory enhances computer performance by preserving frequently used data or instructions so they can be accessed very quickly. During a CPU, the first level cache (L1) is often located inside

P. Das et al. (Eds.): AMRIT 2023, CCIS 1954, pp. 126–138, 2024.
https://doi.org/10.1007/978-3-031-47221-3_12

the processor, whereas the level 3 cache (L3) and the level 2 cache (L2) are situated on different chips.

The two separate components of cache memory are cached data memory and cache tag memory. The numerous collections of memory words that make up cache data memory are referred to as cache blocks, lines, or pages. A block address or tag serves as the identifier for each cache block. A collection of all block addresses or tags is called cache tag memory. The cache memory is accessed when the central processor needs information or instructions by sending an address there.

When data is found in cache, the CPU is informed. Finding data in cache memory is referred to as a cache HIT. When data is not discovered in cache, it is referred to as a cache MISS. In this situation, the address in main memory is searched for the data or instruction. After receiving data from main memory, a block of data is sent from main memory to cache memory, completing all requests from cache. HIT Ratio is a measure of cache memory performance (Fig. 1).

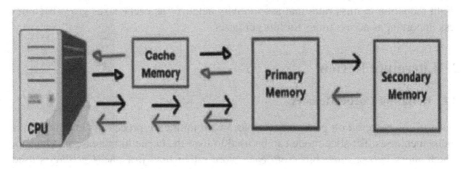

Fig. 1. Block Diagram of Cache Memory

When a software runs on a computer, related storage locations are frequently visited. The temporal segment relates to the use of specific data and/or resources across relatively short time periods. The use of data portions among fairly close storage places is referred to as spatial locality. A mapping technique is used to transport data from main memory to cache memory. Addresses can be mapped to cache locations in three different ways: directly, associatively, and set associatively. Direct mapping, which turns each main memory block into a single possible cache line, is the most basic technique. Associative mapping enables any line of the cache to be mapped into any main memory block. With set associative mapping, any line in a cache set, which is made up of cache lines or blocks, can map any block of main memory (Fig. 2).

The system performance will be impacted by a number of cache design considerations. The line's size is crucial. The level of set associativity is another key trade-off in cache design. On the other hand, increasing the level of set associativity improves the cache's value and quality because it reduces the need for address comparators. Recent studies on cache memory show that, despite the direct mapped cache having a greater miss percentage, on-the-fly mapped caches may routinely beat group associative (or totally associative) caches of equal size. This is brought on by set associative caches'

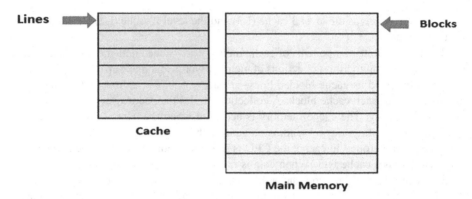

Fig. 2. Cache Mapping

increasing complexity, It considerably increases the access latency for a cache hit. The slight reduction in miss ratio that associativity achieves as cache sizes grow, becomes less important as access times for hits get faster.

2   Literature Review

2.1   Review of Cache Memory

Our work is centered on performance issues with multicore processors and numerous cache memories. Bit slice mode can be used to load the cache to increase parallelism, which slices two or more bits from each word rather than just one. Caching is used to handle the vast majority of memory queries. Cache memories can be successfully implemented in microelectronics memory. Specifies how the computer system should load the cache. Without any explicit programme instruction, the systems can make excellent use of cache. Cache memory has a certain amount of space. Not recommended for high-performance computers. Cached data is only saved for as long as the cache has power [1]. Cache memory serves as a buffer, significantly reducing traffic between the processor and main memory. Cache memory must be used to overcome problems caused by numerous copies of data. Preliminary study indicates that this results in somewhat enhanced speed and less memory congestion; it is either a read or a write miss. Computer programme reuse and locality Cache memory isn't adaptable. It is feasible to assume a hit and proceed in a direct mapped cache. Recover from your miss afterwards. The miss ratio in cache memory is high [2].Cache memory implementation is difficult since each cache access necessitates an inter process call. To avoid congestion, use a bigger cache and more block frames [3]. Cache memory predicts page fault rates reasonably well and provides a good indication of relative performance. It is tractable analytically. Whenever a new block is read into cache, block replacement algorithms are applied (Associative search). No policy for replacing pages exists. Poor performance for cache hits. Their access times are lengthy, and they require varying amounts of electricity while they are on and off [4]. Structured and connected memory is used in the exceptionally fast Josephson NDRO cache design. We intend to eventually put four IK-bit arrays onto a 5.35 × 5.35 mm chip, therefore the discussion has mainly focused on how well they work. Switching is extremely rapid (<10ps). Power dissipation is extremely low (500 nW per circuit). Extremely low temperature operation (=4k). For ultrahigh-performance computers, Josephson devices are appealing. Using zero electrical resistance to conduct strip lines and ground planes. Memory bandwidth and bus traffic memory are also high. Complex circuitry is necessary to evaluate all cache tags in parallel [5].

However, the usage of cache memory has frequently exacerbated rather than reduced the bandwidth problem. Processor memory bandwidth is being reduced. On-chip memory is as quick as a stronger processor. Increasing the hit ratio. Its reducing the time it takes to access data in the cache Keeping multi-cache consistency while minimizing the overheads of updating main memory [6]. On-chip memory must be properly allocated for single-chip computers to attain excellent performance. Pin bandwidth is constrained. It is continuously effective. It takes into account dynamic programme behaviour. It outperforms a data cache in both speed and cost by nearly a factor of two. It could be able to use better techniques and knowledge available at build time to preload data into registers and delete data more efficiently than a data cache. One may consider registers to be local memory. The amount of memory cycles needed is enormous.

Every cache area's register should be updated. Transferring data between memory and cache takes a long time [7]. Each processor's main memory traffic is reduced via cache memory. Transmission between main memory and cache takes less time. If the main memory breadth is narrow, access will take fewer memory cycles. Decrease the amount of address tag bits needed in the cache. Bus protocol should be improved by shortening bus cycles and widening the bus. Replacement plans based on frequency. The cache memory is pretty substantial. The system's performance is most likely to be impacted by cache memory [8]. Miss Caching and Victim Caching to Reduce Conflict Stream buffers and long lines together stream buffers that can prefetch a lot of streams at once. System performance is improved by lowering cache misses. Using 16B lines in stream buffers and victim caches effectively increases the size of the data cache. Only works for tiny caches and has a significant impact on system performance as a whole. It has a large data cache [9]. Increased Write-Back Cache Bandwidth is necessary. Limiting the usage of this code to processors that can execute it and/or have cache line widths less than those expected in the allot directions reduces the write-through cache traffic. Cache allocation directives demand additional processing time. Cache memory also affects high hardware complexity and programme execution time. There is a need for more cycles [10].

Extreme hardware complexity is not employed in high-performance real-time devices. Regarding the ability to access instruction, there is no conflict. Scalability concerns [11] The most frequent technique for bridging the speed difference is cache memory. Cache to increase CPI. Policy of replacement (random, LRU, FIFO) Policy should be written (CBWA, WT). Area is fully associative (Area = PLA + Data + Tag. With each cache, memory must be refreshed. It's possible that syncing caches will cause problems.

It is unable to save the prior state. Pre-fetching is not performed [12]. It is suitable for usage in demanding real-time systems. Reducing the maximum execution time. When the preempted task is reloaded, it returns to its prior state. Worst-case specs, which usually result in processor underutilization. Caches that are extremely huge (up to 511 GB). This has an impact on the architecture of the memory hierarchy's next level and has a negative impact on real-time application performance. The worst-case execution time [13] is considerable. Misses in direct mapped caches are minimized by selective victim caching. It has no impact on cache access time. Data is saved in the main cache. Miss ratio has improved.

Over simple victim caching, increasing the number of interchanges between the main direct-mapped cache and the victim cache. Long access time. When applied to data caches, there is no algorithm. More random data reference patterns are required [14]. Cache memories are short, quick memories that are used to temporarily store memory contents. Caches became an essential component of all processors. When accessing a virtual addressed cache, separable cache, or multiple-access cache, address translation is not necessary. On-chip caches are using up more and more silicon space. Processors are becoming more superscalar. More aggressive designs are required to accommodate several requests per cycle [15].

Choose an experiment to try, such as running commands to determine new locations. Actuate means to send a signal to a device that controls the process (motor, relay etc.) Directly integrated into the application code are inline probes. Problem with bottleneck in embedded systems Memory and high peak power cannot keep up with the CPU [16]. Utilizing several concurrent processes, like in CMP/SMT systems. Lowering interference from cache The processes share the cache in a dynamic manner. a processor with many of useful parts. When processes are being run, large caches are needed. There is a minimum need for extra counters. Performance cannot be improved if caches are too small for the needs [17]. Utilize memory more efficiently by using a 3D graphics cache.. Resolving a memory bottleneck issue in an embedded system Reduce the number of intense memory accesses in a short period of time while increasing memory utilization. Supporting an AXI low power interface reduces power usage. Enhancing the on-chip bus. Implementation is challenging [18]. Multicore processor with high speed. To make the timing better. We reduce control and clock loads by aligning cell instances into "vectors." The Most Significant Advantage Quick area and timing feedback to the DE. Different data path configurations can be quickly prototyped. The replacement circuit gets more complicated. Complexity is high. Cache memory design is difficult [19]. Police investigation cache miss design using FPGA cache memory lower power usage and improved performance. The induced miss rate in cache memory is monitored by the cache controller. On-chip power usage. There aren't any substitute policies. The processing time is lengthy [20] (Fig. 3).

Faster replacement methods often track less usage data. To shorten the time it takes to update that information. Regarding the direct-mapped cache, there is no information. Apps for streaming audio and video Reusable data should be cached (cache pollution). Nearly one billion transistors are included on a single device. The full capacity of world scientific codes has already been reached. Scientific computing often necessitates memory ranging from a few Megabytes to hundreds of Gigabytes [21]. By utilizing greater block size, larger cache, and better prediction algorithms, the rate of conflict misses is minimized. However, using a larger block size may result in a higher miss penalty, shorter hit time, and higher power usage. Larger cache, on the other hand, results in slow access

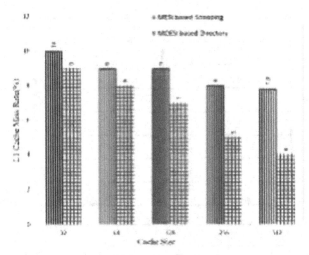

Fig. 3. Cache Size

time and expensive cost. It is related to the cache coherence issue. Higher associativity produces shorter intervals but requires a shorter cycle duration. Matrix multiplication of various sizes is a problem. It is difficult to pinpoint a specific cache optimization strategy. Larger caches result in slower access times and higher costs [22]. Higher associatively produce shorter intervals, but they require a shorter cycle period. When compared to Cache miss look aside, The victim cache minimizes the miss rate at a significant expense. Overall, we can claim that there is a huge opportunity to improve cache memory performance. Cache Memory influences software execution time [23]. Flash memory offers excellent performance, large capacity, and consistent service quality.

Increases the amount of data that the CPU can handle. Cache size was increased from 3KB to 15KB, with a 273.8% increase in random write speed and a 214.4% increase in random read performance. The amount of traffic between the processor and the cache memory grows. Cache misses result in a cache miss penalty, which reduces total storage performance [24]. Cache performance based on replacement rules such as LRU, FIFO, and Random. They discovered that the LRU policy in the data cache outperforms FIFO and Random. The replacement policies for instruction caches. They put forward a new loop model. In its loop model, they discovered that random replacement outperformed LRU and FIFO. However, each simulation has a distinct cache size, cache associativity, and benchmarks. As a result, policy performance comparisons are less reliable. To compare the performance of all policies, a single simulation should be constructed [25] (Fig. 4).

2.2 Review of On-Chip Cache Architecture to Reduce Latency

In [26] Kim proposed the physical design for Non-Uniform Cache architecture (NUCA). Kim considered single core processor to design the NUCA. The performance of cache can be improved by increasing the number of banks which reduce the channel contention. Initially Kim implemented a design with dedicated channel to bank so that routing

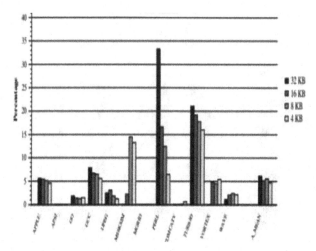

Fig. 4. Traffic reduction

overhead can be reduced. But private per-bank channel increase the wire and decoder overhead on the chip. To avoid this wire overhead, two-dimensional on-chip network is designed to interconnect all the banks as shown in Fig. 5(a). These types of cache are called Static-NUCA (SNUCA) as the memory blocks remain in the fixed banks. The access latency can be improved by transferring frequently accessed blocks closer to the core as the bank in the close proximity are faster than other. Such design is termed as Dynamic-NUCA (DNUCA).

To achieve this dynamic behavior of cache architecture, three basic policies are followed: (i) mapping: in which bank the data should reside and how to allocate this bank, (ii) search: how to find out a data block in a bank and (iii) movement: under what condition a block should change the location, how to change the location and where to place. In SNUCA, a block can be mapped in any predefined bank or in any bank of the cache as shown in Fig. 5(b). To locate any data block all the banks must be searched. To reduce this searching cost associative structure is implemented where multiple bank lie in a set and each bank occupy one way. Ex. in a 4-way cache, each set consists of 4 banks. There are three basic mapping techniques: (i) simple mapping (ii) fair mapping (iii) shared mapping. In simple mapping, a data block can be searched by first selecting the set and then performing the tag match within this set. The main drawback of this mapping is that some set will be always in the close proximity to the core resulting faster access compared to other. In fair mapping the first way of a set is accommodated close to the core so that average access latency for each bank set remain same. Shared mapping achieve fastest access to all bank by sharing the closest bank to the core. To locate a data block searching can be done in three ways: (i) incremental, (ii) multicast and (iii) combined. In incremental searching technique, the search starts from the closest bank and subsequently search one after another. It is simple to implement and also network latency is less. In multicast search, all the bank sets are searched simultaneously. It is faster than the incremental but network communication overhead is higher due to deployment of routing technique. Even though it is a parallel search but physically one bank sets is searched at a time. Combined search locate a bank by performing multicast search over some banks and remaining banks are searched in incremental manner. Kim proposed a smart search technique by adding partial tagging mechanism. A partial tag will be stored for each block so that on request of a data block, this partial tag will select a subset of banks where the data block is likely to be present.

Kim proposed a simple block movement mechanism to (a) (b) migrate one frequently accessed data block closer to the core. His Generation Promotion mechanism places a newly fetched block into the farthest way. Upon every access, the block is interchanged with the adjacent bank closer to the cache controller.

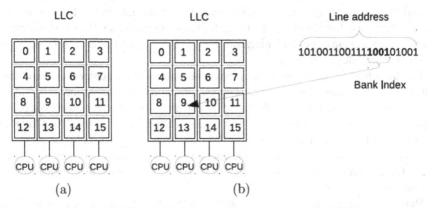

Fig. 5. (a) NUCA with 16 banks (b) bank mapping in SNUCA.

Kim's proposal of NUCA uses data block mapping with the help of partial tagging technique for single core processor. Also allowed migration from the farthest location to the nearest location via high bandwidth network.

Chishti et al. [27] introduces the distance associativity concept to improve the sequential access to the data block and migration with few places swapping. The architecture is named as Non Uniform access with Replacement And Placement using Distance associativity (NuRAPID). Distance associativity divides the datasets into multiple groups at die rent distance having die rent access latency. In this design, tag array is placed close to the core and is accessed in sequential manner. To access a data block, the tag array is searched first to second the exact location of bank in which the data reside. By separating tag array from data array, it is possible to place data block of same set in the same d-group or in different d-groups. So, a newly fetched data block can be store in any set of fastest d-group without demoting any block to the farthest d-group.

Upon access to a block, tag array returns a pointer to the exact location of data block in a particular dataset. Beckman and wood [28] proposed the design of multicore NUCA architecture. He considered the basic changes to design large cache architecture for Chip Multicore Processor (CMP). Multicore processor shares the last level cache requiring fast communication between cache and core. Different process works with different dataset compete with each other to utilize the common LLC. Also, fastest bank for one core may be the slowest bank for other core. In this paper, Beckman CMP of eight core associated with private L1 surrounded by the shared large LLC. Here, minimum bank size is chosen to reduce network bandwidth requirement. A large with link is designed to connect each router of four bank to speed up the communication. The large shared cache is divided into 16 Tetris shaped zone, each consists of 16bank from 16 different bank sets. 16 Tetris zones are further categorized into three parts. The 8 Tetris zone close to 8 core form the local region as each Tetris is very close to one core. The 4 Tetris zones residing at central region of the cache are called center and the remaining 4 Tetris are called inter or intermediate region. Beckman block migration involves three policies: (i) allocation, (ii) migration and (iii) searching. Initial placement follows a static allocation by considering a lower order bit of each tag to select a particular bank within a bank-cluster. The intention of a migration policy is to maximizing the L2 hit in the local region of a core. To achieve this, if a requesting block is directly migrated to the local cluster can increase the access latency of a distant core for the same block. So, Beckman migration policy follows a chain of gradual migration as Otherlocal-> otherinter- > othercenter-> mycenter-> myinter-> mylocal.

By this policy, equally shared data blocks concentrate in the center region and inter contain the blocks shared by the adjacent cores. This architecture implements two phase multicast searching. In first phase, 6 Tetris closely associated with a core will be accessed by passing message to all the banks. These 6 Tetris includes one my local, one my inter and 4 center. This cluster of Tetris have the maximum chance to hold the requested block as the migration policy moves the data block towards the local region on every access. On request miss in _rst phase, the rest 10 Tetris are searched by broadcasting the message.

Huh et al. [29] organizes the L2 cache into two broad categories (i) shared and (ii) unshared. Unshared region of cache allocates a private section of L2 cache to each core where shared region is made available to all the core on the chip. An optimal degree of sharing policy is proposed to divide the cache into these two regions based on applications. Cache miss can be reduced by sharing cache among maximum number of core. Greater degree of sharing will allow to store maximum blocks shared by multiple cores resulting in reduced cache miss. But it increases the access latency by keeping maximum blocks far from the core. Sharing degree directly elects the hit latency, hit rate, process to process communication and difficulties in coherence maintenance. Hit latency decreases with lowering the degree of sharing as few shared block will be at remote location. With higher degree of sharing, hit rate improves as a core can access maximum cache space and also a core shares maximum blocks with other core resulting improved inter-process communication. But, it suffers in maintaining the L1 coherence as L2 cache can be accessed and update by other cores. Huh experimentally observe the performance of this cache architecture with different degree of sharing on various commercial applications and scientific benchmarks. Huh shown that SNUCA architecture with 2 or 4 degree of sharing yields better performance.

Liu et al. [30] have introduces Shared Processor based Split L2 cache that takes the maximum advantages of shared and unshared L2 design and avoids the drawbacks. Private 12 cache performs better when multiple core shares few blocks of memory. In this design a bus is employed instead of interconnection network and L1 cache is placed on the core side of bus. The shared L2 cache splits are places on the other side of bus. Each split of L2 cache is assigned to one core. Upon a L1 cache miss, request is passed to common bus to access the split(s) assigned to the core which has generated the request. Even if the block is not available in the assigned splits, request is passed to other splits before accessing o_ chip memory.

This lookup policy of L2 can be classified into three cases:

1. Local hit: if the data block is available in the assigned L2 split(s).
2. Remote hit: if the data block is found in the neighbor L2 splits
3. L2 miss: if data block is not available in any of the L2 split, the request is passed to on-chip memory.

If a CPU request is responded by a split other than the assigned split(s), the data block will be placed in an assigned split. If the data block is not found in any L2 split, then the data block will be fetched from on-chip memory and will be placed in one of the assigned split.

Dybdahl et al. [31] proposed a NUCA architecture which combines the concept of private LLC cache and shared LLC cache. In this design, three level of cache is considered where L3 is the last level cache. Each core is in close proximity to a portion of shared L3 cache, access to which is faster compared to other portion of L3. An adaptive scheme is employed on the L3 cache to use a part of L3 cache as shareable memory. Each set of shared cache is distributed across all the cores. On frequent access, a block is shifted towards the private portion of L3. For any request, the tag of a block is searched in private portion first then, subsequently search the shared portion. Dybdahl proposed a dynamic L3 cache partitioning by allocating variable number of ways to private partition. Each set maintained a shadow tag for each core. To evict a cache block from L3, it should be recorded in shadow tag of core which fetched it. On next cache miss, the shadow tag for that set is searched first, if it is found, then the counter hit in shadow tag for requesting core is increased by one.

SP-NUCA architecture is a simple hybrid architecture proposed by Merino et al. [32]. In this design, LLC is divided into multiple sets where each set can store private and shared block of data. Private blocks are assigned to a particular core and shared blocks are made available to all the core. To search any block, the tag is compared within the private blocks of the requesting core. If there is no match then, the shared blocks are accessed. If there is a miss in shared block too, the private blocks assigned to other will be searched. If a data block is found in private block of other core, it will be migrated to the shared blocks. Initially, a data block fetched from the memory is placed in this private block of fetching core. The migration policy is implemented with an additional private bit. By piping the bit, a block is reassigned as private or shared without swapping the data block physically.

In 2009, Madan et al. [34] proposed a 3D hybrid cache design by stacking SRAM and DRAM upon Chip Multicore Processor to increase the cache capacity. Their aim is to take the advantages of low latency SRAM and high density of DRAM effectively. They have also reduces wire latency by stacking three layers, where top two layers consists of 16 DRAM banks upon 16 SRAM banks act as L2 cache. Request from the 16 cores of bottom layer reaches SRAM layer first then passes to DRAM layer. In 2011, Inoue et al. [35] have improve the cache performance by proposing an SRAM/DRAM hybrid cache architecture consists of small sized SRAM and large stacked DRAM.

The cache operates in two modes depending on the demand of program. If a program requires large capacity cache, then DRAM mode uses large DRAM while small SRAM is used store the tags only. In 2013 [36] Hameed et al. observed that the hierarchy cache consists of L3 SRAM and L4 DRAM, supers with redundant data. So, they proposed a policy for hybrid SRAM-DRAM LLC to decide whether a new block is to be placed in both SRAM and DRAM or only in DRAM to avoids such data duplication. Later in 2014, Hameed et al. [37] proposed a technique to avoid the high tag latency of large DRAM cache by designing two small SRAM to hold the tags of frequently referred sets of large L3 SRAM and large L4 DRAM separately. They named these small SRAMs as SRAM tag-cache and DRAM tag-cache. Huang et al. in 2014 [38], proposed a technique to avoid the area overhead of tags-in-SRAM compared to tags-in DRAM without degrading the performance. They have used DRAM to maintain tags along with a small dedicated SRAM cache, called aggressive tag-cache (ATCache) that caches the recently accessed tags to utilize temporal locality. Gulur et al. in 2015 [39], proposed an adaptive performance model of the DRAM Last Level Cache that spans both tags-in-SRAM and tags-in- DRAM organizations. Their model estimates the average access latency and request arrival rate at DRAM cache.

3 Conclusion

We addressed cache memory and how to increase cache memory performance using all of the references in this paper. The central processing unit of the system uses cache memory, which is a fast random-access memory, to temporarily store data and instructions. By keeping the most frequent and likely data and instructions "near" to the processor, where the system central processing unit may readily recognize them, it decreases execution time. The data structures used today will be replaced by larger, more intelligent caches in the future. The RAM requirements for today's scientific computing applications range from several Megabytes to many Gigabytes.

References

1. Curtis, H.A.: Systematic procedures for realizing synchronous sequential machines using flip-flop memory: part I. IEEE Trans. Comput. **100**(12), 1121–1127 (1969)
2. Winder, R.0.: A data base for computer performance evaluation. In: Presented at IEEE Workshop on System Performance Evaluation, Argonne, Ill., October 1971, and published in COMPUTER, vol. 6, no. 3 (1973)
3. Chow, C.K.: An optimization of memory hierarchies by geometric programming. In Proceedings of 7th Annual Princeton Conference Information Science Systp. vol. 15 (1973)
4. Aven, O.I., Boguslavsky, L.B., Kogan, Y.A.: Some results on distribution-free analysis of paging algorithms. IEEE Trans. Comput. **25**(07), 737–745 (1976)
5. Gueret, P., Moser, A., Wolf, P.: IBM J. Res Develop. 24 (1980, this issue)
6. Amdahl, C.: [Amdahl 82] private commuvtict~tiovt, March 82
7. Hasegawa, M., Shigei, Y.: High-speed top-of-stack scheme for VLSI processor: a management algorithm and its analysis. ACM SIGARCH Comput. Architect. News **13**(3), 48–54 (1985)
8. Winsor, D.C., Mudge, T.N.: Analysis of bus hierarchies for multiprocessors. ACM SIGARCH Comput. Architect. News **16**(2), 100–107 (1988)

9. Baer, J.L., Wang, W.-H.: On the inclusion properties for multi-level cache hierarchies. In The 15th Annual Symposium on Computer Architecture, pp. 73–80. IEEE Computer Society Press, June (1988)
10. Smith, A.J.: Second bibliography on cache memories. ACM SIGARCH Comput. Architect. News **19**(4), 154–182 (1991)
11. Bakoglu, H.B., Grohoski, G.F., Montoye, R.K.: The IBM RISC system/6000 processor: hardware overview. IBM J. Res. Dev. **34**(1), 12–22 (1990)
12. Flynn, M.J.: Computer Architecture: Concurrent and Parallel Processor Design (1994)
13. Lebeck, A.R., Wood, D.A.: Cache profiling and the SPEC benchmarks: a case study. Computer **27**(10), 15–26 (1994)
14. Stiliadis, D., Varma, A.: Selective victim caching: a method to improve the performance of direct-mapped caches. IEEE Trans. Comput. **46**(5), 603–610 (1997)
15. Peir, J.K., Lee, Y., Hsu, W.W.: Capturing dynamic memory reference behavior with adaptive cache topology. In: Proceedings of the Eighth International Conference on Architectural Support for Programming Languages and Operating Systems, pp. 240–250 (1998)
16. Sebek, F., Gustafsson, J.: Determining the worst case instruction cache miss-ratio. In: Proceedings of Workshop On Embedded System Codesign (ESCODES'02) (2002)
17. Yang, S.H., Powell, M.D., Falsafi, B., Vijaykumar, T.N.: Exploiting choice in resizable cache design to optimize deep-submicron processor energy-delay. In: Proceedings Eighth International Symposium on High Performance Computer Architecture, pp. 151–161. IEEE (2002)
18. Lee, K.W., Park, W.C., Kim, I.S., Han, T.D.: A pixel cache architecture with selective placement scheme based on z-test result. Microprocess. Microsyst. **29**(1), 41–46 (2005)
19. Temam, O.: An algorithm for optimally exploiting spatial and temporal locality in upper memory levels. IEEE Trans. Comput. **48**(2), 150–158 (1999)
20. Almoosa, N., Wardi, Y., Yalamanchili, S.: Controller design for tracking induced miss-rates in cache memories. In: IEEE ICCA 2010, pp. 1355–1359. IEEE June (2010)
21. Ahmed, N. Mateev, N., Pingali, K.: Tiling imperfectly nested loop nests. In: Proceedings of the ACM/IEEE Supercomputing Conference, Dallas, Texas, USA (2000)
22. Xu, Y., et al.: A novel cache size optimization scheme based on manifold learning in content centric networking. J. Netw. Comput. Appl. **37**, 273–281 (2014)
23. Kampe, M., Stenstrom, P., Dubois, M.: Self-correcting LRU replacement policies. In: Proceedings of the 1st Conference on Computing Frontiers, pp. 181–191 (2004)
24. Xilinx, UG585 Zynq-7000 All Programmable SoC Technical Reference Manual (v1.11) (2016)
25. Burger, D., Austin, T.M.: The SimpleScalar tool set, version 2.0. ACM Sigarch Comput. Architect News **25**(3), 13–25 (1997)
26. Kim, C., Burger, D., Keckler, S.W.: An adaptive, non-uniform cache structure for wire-delay dominated on-chip caches. In: Proceedings of the 10th International Conference on Architectural Support for Programming Languages and Operating Systems, pp. 211–222 (2002)
27. Chishti, Z., Powell, M.D., Vijaykumar, T.N.: Distance associativity for high-performance energy-efficient non-uniform cache architectures. In: Proceedings. 36th Annual IEEE/ACM International Symposium on Microarchitecture, 2003. MICRO-36, pp. 55–66. IEEE (2003)
28. Beckmann, B.M., Wood, D.A.: Managing wire delay in large chip-multi processor caches. In: 37th International Symposium on Microarchitecture (MICRO-37'04), pp. 319–330. IEEE (2004)
29. Huh, J., Kim, C., Shafi, H., Zhang, L., Burger, D., Keckler, S.W.: A NUCA substrate for flexible CMP cache sharing. In: ACM International Conference on Supercomputing 25th Anniversary Volume, pp. 380–389 (2005)

30. Liu, C., Sivasubramaniam, A., Kandemir, M.: Organizing the last line of defense before hitting the memory wall for CMPs. In: 10th International Symposium on High Performance Computer Architecture (HPC 2004), pp. 176–185. IEEE (2004)
31. Dybdahl, H., Stenstrom, P.: An adaptive shared/private NUCA cache partitioning scheme for chip multiprocessors. In: 2007 IEEE 13th International Symposium on High Performance Computer Architecture, pp. 2–12. IEEE (2007)
32. Merino, J., Puente, V., Prieto, P., Gregorio, J.Á.: SP-NUCA: a cost effective dynamic non-uniform cache architecture. ACM SIGARCH Comput. Architect. News **36**(2), 64–71 (2008)
33. Mittal, S., Vetter, J.S., Li, D.: A survey of architectural approaches for managing embedded DRAM and non-volatile on-chip caches. IEEE Trans. Parallel Distrib. Syst. **26**(6), 1524–1537 (2014)
34. Madan, N., et al.: Optimizing communication and capacity in a 3D stacked reconfigurable cache hierarchy. In 2009 IEEE 15th International Symposium on High Performance Computer Architecture, pp. 262–274. IEEE (2009)
35. Inoue, K., Hashiguchi, S., Ueno, S., Fukumoto, N., Murakami, K.: 3D implemented SRAM/DRAM hybrid cache architecture for high-performance and low power consumption. In: 2011 IEEE 54th International Midwest Symposium on Circuits and Systems (MWSCAS), pp. 1–4. IEEE (2011)
36. Hameed, F., Bauer, L., Henkel, J.: Adaptive cache management for a combined SRAM and DRAM cache hierarchy for multi-cores. In: 2013 Design, Automation & Test in Europe Conference & Exhibition (DATE), pp. 77–82. IEEE (2013)
37. Hameed, F., Bauer, L., Henkel, J.: Reducing latency in an SRAM/DRAM cache hierarchy via a novel tag-cache architecture. In: Proceedings of the 51st Annual Design Automation Conference, pp. 1–6 (2014)
38. Huang, C.C., Nagarajan, V.: ATCache: Reducing DRAM cache latency via a small SRAM tag cache. In: Proceedings of the 23rd International Conference on Parallel Architectures and Compilation, pp. 51–60 (2014)
39. Gulur, N., Mehendale, M., Govindarajan, R.: A comprehensive analytical performance model of dram caches. In: Proceedings of the 6th ACM/SPEC International Conference on Performance Engineering, pp. 157–168 (2015)
40. Balasubramanian, R., Jouppi, N.P., Muralimanohar, N.: Multi-core cache hierarchies. Synth. Lect. Comput. Architect. **6**(3), 1 (2011)

Authenticating Smartphone Users Continuously Even if the Smartphone is in the User's Pocket

Sandip Dutta[1]([⊠]) [iD], Soumen Roy[2], and Utpal Roy[3]

[1] Department of Computer and System Sciences, Visva-Bharati, Santiniketan, West Bengal 731235, India
03333342105@visva-bharati.ac.in
[2] Department of Computer Science and Engineering, University of Calcutta, Acharya Prafulla Chandra Roy Siksha Prangan, JD - 2, Sector - III, Saltlake City, Kolkata, West Bengal 7000106, India
[3] Department of Computer and System Sciences, Visva-Bharati, Santiniketan, West Bengal 731235, India

Abstract. Users often underestimate the power and utility of smartphones, as they are unaware of the wide range of intelligent sensors that are now built-in into them. Smartphones are equipped with a wide array of sensors, including accelerometers, gyroscopes, magnetometers, barometers, pedometers, etc. These sensors are continuously active whenever the phone is turned on. Our study examines the way in which a smartphone can authenticate a user in a continuous and automatic manner, without requiring their awareness.

To achieve this goal, we have created a moderate-sized human gait database that comprises data from 63 volunteers, taking into account their ages, genders, educational levels, professions, and backgrounds. In order to create this database, a web-based application was used to utilize the sensors on smartphones to capture human gaits. To determine which algorithm is most effective at verifying users based on their gait patterns, we analyzed the dataset here using ten popularly known anomaly detection algorithms.

We found that one-class support vector machines (OCSVMs) are the most powerful for verifying users based on gait, with an average Equal Error Rate (EER) of 0.0043. It has been also identified that the EER has been reduced from 0.0043 to 0.0013 when soft biometric traits like gender, education level, and age group information are incorporated. We have proposed research directions to develop and implement this technology, and we believe it has practical applications.

Keywords: Smartphone Sensors · Gait Analysis · Anomaly Detection · One-Class SVM

1 Introduction

In the recent past, S. Roy et al. [1–10] and his group has published number of interesting article for identification, verification and authentication purposes by analyzing the smartphone sensors data collected appropriately through self-developed mobile application

P. Das et al. (Eds.): AMRIT 2023, CCIS 1954, pp. 139–146, 2024.
https://doi.org/10.1007/978-3-031-47221-3_13

software. In their study [10], human gait has been used for identification and verification purposes. This study has been extended for the continuous identification and verification of the user through smartphone data. It indicates that the intelligent sensors of the smartphone will continuously authenticate and verify only the user of it by capturing the sensor data. This verification is related not only through the gait, but it also incorporates the body language of the user too.

1.1 Motivation and Opportunities

The usage of mobile technology has rapidly expanded across the world. Presently, it is approximated that more than 5 billion individuals possess mobile devices [16], with more than half of these connections being smartphones. Smartphones come with several advanced sensors [10] that allow them to gather a variety of detailed contextual data, including information on location, device utilization, human movement pattern, etc. at a particular moment. The effective utilization of the information has the potential to provide us with valuable insights and may be considered as a promising solution for many upcoming technologies like analysingbehaviouralbiometrictraits,E-healthcareetc.Unfortunately,thevaluable information produced by smartphone sensors has not yet been fully utilized to its potential. This suggests that there is a significant amount of untapped potential for further research and development utilizing smartphone sensors.

1.2 Literature Survey

The term "human gait" refers to how a person walks, specifically the style or pattern that is unique to each individual. Traditionally, gait analysis has primarily been employed for medical assessment objectives by using video images to monitor a patient's bodily movements or body language. However, this approach is costly and requires the use of an indoor laboratory. As a result, Morris et al. [11] introduced a revolutionary approach in 1973 that entailed the insertion of an accelerometer sensor into the human body for the purpose of identifying bodily movements. However, the challenge with information gathered solely through an accelerometer sensor is that inaccuracies escalate with velocity. As a result, to mitigate this undesirable error, Dejnabadi et al. 2006 [12] as well as Favre et al. in 2008 [13] utilized a fusion of accelerometer and gyroscope sensors. Several subsequent investigations have adopted a similar technique as the usage of a fusion of sensors leads to precise identification of the user. But these research is founded on the application of external sensors affixed to the human body, which still encounters issues with expense as well as information is limited to three to four sensors only.

In the past, limited research focused on using smartphone sensors to track human gaits. But it is becoming a recent research trend to its easy availability, accurate information, and cost-effectiveness. A recent study [14] suggested a method for authenticating smartphone users based on their behavioural characteristics, utilizing smartphone sensors. The researchers named their system "SmarterYou", and through the use of the kernel ridge regression machine learning technique, they have achieved an accuracy rate of 98% in user authentication. Another recent study [15] introduced a plan that involved gathering information from various 3D sensors on a smartphone in the background over a set period and using the collected hand movements to create a profile of the user. They have achieved a True Acceptance Rate (TAR) of 96% with an Equal Error Rate (EER) of 4% when testing their approach on a group of 31 qualified volunteers using the Random Forest (RF) classifier.

Figure 1 presented below shows the current level of research conducted around the world on gait analysis.

1	Mozambique	100
2	Angola	61
3	Brazil	56
4	Portugal	56
5	United Kingdom	19

Fig.1. Human gait analysis interest worldwide

2 Major Goals and Contributions

The paper has four main goals and contributions, which are outlined as follows.

1. The first aim is to introduce a framework for creating a dataset on gait analysis by using sensors on smartphones while walking.
2. The second objective is to suggest a method for continuous user authentication by utilizing an anomaly detector based on OCSVM.
3. The third objective is to use multiple soft biometric attributes to boost the effectiveness of the proposed model suggested in the paper [2].
4. The fourth aim is to compare the performance of the proposed anomaly detector with various advanced and contemporary detectors.

3 Methodology

3.1 Dataset Preparation

- To prepare the dataset, 63 volunteers were involved, including individuals with varying professions, genders, age groups, and educational backgrounds. Figure 1 shows the class distribution of the subjects. Samples have been collected with the assistance of a Samsung M20 android mobile device.
- From each participating individual data has been collected over the course of 5 sessions, and each session included four repetitions, by a web-based application developed by us. There was a 12-h interval between the two sessions.
- At the time of collecting the samples of individuals, the smartphone holding the web-based application has been placed at the pant pocket location of the volunteer. The volunteer was then asked to walk on a flat surface for a duration of 1 min and 30 s in a usual manner.

- As the volunteer moved, the smartphone placed in their pocket started to shift position, and the web-based application records the sensory information produced by changing the position of the smartphone. This way, during each session, four samples were collected from each volunteer.
- In addition to the acceleration, rotation, gyroscopic, and gravitational information of the smartphone generated due to the gait from the sensors; we have also gathered information about volunteer traits to use a soft biometric method [2] and demonstrate how these traits can affect the results (Fig. 2).

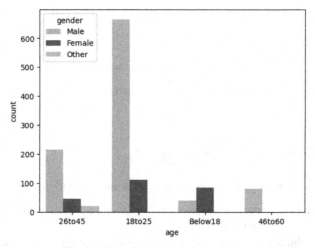

Fig. 2. Class distribution of the subjects

3.2   Model Implementation

The actions to carry out the implementation are outlined as:

- Our team has developed a Shinyapp using the R statistical programming language to execute anomaly detection algorithms on our dataset.
- The data has been organized in a.csv format within an Excel file.
- The dataset has been uploaded to the Shinyapp and employed 10 popular anomaly detection algorithms to analyse it.
- To facilitate training, the dataset has been split into two sessions, and three sessions have been allocated for testing purposes.
- We have incorporated several soft biometric traits using the Shinyapp and observed the impact of their inclusion on the results obtained from the anomaly detection algorithm.

The following ten anomaly detection algorithms were used to analyze the dataset: OCSVM, Scaled-Manhattan, Hamming, Bhattacharyya, Euclidean, Czekanaowski, Squared_Chord, Mean, Motkya, Dice.

Several soft biometric characteristics that have been considered include Gender, Age Groups, Qualification, Handedness, Hands Used, and Chronological Information.

4 Results and Discussions

4.1 Detector Performance Evaluation Metrics

The user authentication model's performance is often assessed using a commonly used metric called the Average Equal Error Rate (EER). This metric involves setting a threshold for acceptance, at which the False Acceptance Rate (FAR) and False Rejection Rate (FRR) are equal. The False Acceptance Rate (FAR) represents the proportion of unauthorized individuals who are wrongly granted access, out of the total number of attempts made by those individuals to gain access. The less value of False Acceptance (FAR) indicates a higher level of security. The False Rejection Rate (FRR) is the ratio of the number of unsuccessful verification attempts made by a legitimate user to the total number of verification attempts made by that user. A higher False Rejection Rate (FRR) is typically associated with a lower usability rating.

When it comes to the user authentication model, setting a higher threshold tends to decrease the False Acceptance Rate (FAR) while increasing the False Rejection Rate (FRR). As a result, selecting an appropriate threshold value is a crucial factor in reducing the Equal Error Rate (ERR).Our study emphasizes the importance of carefully evaluating both security and usability requirements when selecting a threshold for each individual user, in order to ensure an appropriate balance.

4.2 Performance Comparison

The prepared gait database has been analyzed with R-studio and obtained results are placed in Table 1.

Table 1. Performance comparison of the anomaly detectors.

Anomaly Detectors	Without Soft Biometric	With Soft Biometric Gender	Plus Age Group	Plus Handedness	Plus Qualification	Plus Hands Used	Plus Chronological Information
One-Class SVM	**0.0043**	**0.0045**	**0.0040**	**0.0034**	**0.0023**	**0.0015**	**0.0013**
Scaled-Manhattan	0.0437	0.0437	0.0435	0.0435	0.0431	0.0429	0.0429
Hamming	0.1023	0.1021	0.1012	0.1009	0.0993	0.0987	0.0987
Manhattan	0.0918	0.0914	0.0908	0.0896	0.0879	0.0862	0.0862
Bhattacharyya	0.0914	0.0907	0.0882	0.0853	0.0816	0.0767	0.0761
Euclidean	0.0897	0.0892	0.1189	0.1155	0.1114	0.1071	0.1066
Czekanaowski	0.086	0.0859	0.0851	0.0845	0.0829	0.0810	0.0810
Squared_Chord	0.1345	0.1314	0.1266	0.1215	0.1161	0.1097	0.1097
Motkya	0.1146	0.1142	0.1129	0.1115	0.1094	0.1074	0.1073
Dice	0.1266	0.1254	0.1221	0.1177	0.1139	0.1085	0.1084

Here, each numerical value indicates the average equal error rate (EER) that was achieved by using the corresponding anomaly detection algorithm on our gait dataset. We have considered 10 popularly known anomaly detectors as mentioned in the Table 1 to evaluate the Equal Error Rates. The various soft biometric traits have been incorporated one by one in the input of the anomaly detection algorithm and the performance through EER has been observed. The result of each column of the table indicates the improved EER with the inclusion of separate soft biometrics with the earlier result. So it can be concluded that the recognition rate will increase gradually with the incorporation of various soft biometric traits and will reach a maximum value.

It is evident from Table 1 that the OCSVM achieves the lowest average equal error rate for user identification based on their gait pattern, regardless of whether a soft biometric approach is used or not. As a result, we have chosen the OCSVM as the preferred model for our system.

If this application continuously runs within the user's mobile, the machine will be more and more intelligent in course of time to identify and verify the particular user as well as the owner of it. If the mobile set goes to a user other than the owner, it will give some indication and may be equipped to switch off, mentioning unauthorized access.

4.3   Major Contribution of the Study

The present study report a powerful and cost-efficient identification and authentication technique analyzing human gaits.

- Acontinuoususerauthenticationtechniquehasbeenproposedbyconducting the experiment with ten different anomaly detection algorithms. Considering the performance; in this study, OCSVM has been considered as the best one.
- The nature of EER has been examined with the incorporation of various soft biometric characteristics, and it has been found that the OCSVM algorithm outperforms all others considered here.

5   Conclusion

Smartphone sensors can collect data beyond the conscious awareness of users that can be used to authenticate them beyond their conscious awareness. This technology holds substantial implications for security and privacy, particularly in its applications within the medical insurance industry. The technology's capability to perform continuous authentication of smartphone users, irrespective of their device's location within their pockets, presents a powerful solution for identity verification. Furthermore, it has potential for use in safeguarding sensitive health-related information for the purpose of granting discounts in new health insurance campaigns. By continuously verifying the identity of users, the technology can help to reduce the risk of fraudulent activities, including falsifying insurance claims or utilizing others' insurance policies. This feature can be especially beneficial in the context of the medical insurance industry, where ensuring the security and privacy of sensitive medical data is of paramount importance.

Acknowledgements. We express our profound gratitude to DST-INSPIRE for their financial support and contributions, without which the fulfillment of this research project would not have been achievable.

References

1. Roy, S., et al.: A systematic literature review on latest keystroke dynamics based models. IEEE Access **10**, 92192–92236 (2022). https://doi.org/10.1109/ACCESS.2022.3197756
2. Roy, S., Roy, U., Sinha, D.D.: Efficacy of typing pattern analysis in identifying soft biometric information and its impact in user recognition. In: Battiato, S., Farinella, G.M., Leo, M., Gallo, G. (eds.) ICIAP 2017. LNCS, vol. 10590, pp. 320–330. Springer, Cham (2017). https://doi.org/10.1007/978-3-319-70742-6_30
3. Roy, S., Roy, U., Sinha, D.D.: Enhanced knowledge-based user authentication technique via keystroke dynamics. Int. J. Eng. Sci. Invention (IJESI). **3**(9), 41-48 (2014)
4. Roy, S., Roy, U., Sinha, D.D.: Password recovery mechanism based on keystroke dynamics. In: Mandal, J.K., Satapathy, S.C., Sanyal, M.K., Sarkar, P.P., Mukhopadhyay, A. (eds.) Information Systems Design and Intelligent Applications. AISC, vol. 339, pp. 245–257. Springer, New Delhi (2015). https://doi.org/10.1007/978-81-322-2250-7_24
5. Roy, S., Roy, U., Sinha, D.D.: Rhythmic password-based cryptosystem. In: 2nd International Conf. on Computing and System, pp. 303–307, University of Burdwan, West Bengal, India (2013). ISSN 2194–5357, ISBN 978–81322–2249–1
6. Roy, S., Roy, U., Sinha, D.D.: Performance perspective of different classifiers on different keystroke datasets. Int. J. New Technol. Sci. Eng. (IJNTSE) **02**(04), 64–73 (2015). ISSN: 2349–0780
7. Roy, S., Roy, U., Sinha, D.D.: Distance-based models of keystroke dynamics user authentication. Int. J. Adv. Eng. Res. Sci. (IJAERS) **02**(09), 89–94 (2015). ISSN: 2349–6495
8. Roy, S., Roy, U., Sinha, D.D.: Modified knowledge-based user authentication technique. In: 7th International Conference on Mathematical Science for Advancement of Science and Technology, MSAST, IMBIC, Kolkata, India, vol. 2 (2013)
9. Roy, S., Roy, U., Sinha, D.D.: Combined user authentication technique. In: International Conference on Recent Trends in Science Technology (ICRTST), College of Engineering and Management, Kolaghat, pp. 106–113, West Bengal, India (2013)
10. Ghosh, D., Roy, S., Roy, U.: Gait identity verification using equipped smartphone sensors. In: National Conference on Emerging Trends on Sustainable Technology and Engineering Applications (NCETSTEA) (2020). ISBN: 978–1–7281–4362–0, https://doi.org/10.1109/NCETST EA48365.2020.9119955
11. Morris, J.R.W.: Accelerometry—a technique for the measurement of human body movements. J. Biomech. **6**, 729–736 (1973)
12. Dejnabadi, H., Jolles, B.M., Casanova, E., Aminian, P.F.K.: Estimation and visualization of sagittal kinematics of lower limbs orientation using body-fixed sensors. IEEE Trans. Biomed. Eng. 1385–1393 (2006). https://doi.org/10.1109/TBME.2006.873678
13. Favre, J., Jolles, B.M., Aissaoui, R., Aminian, K.: Ambulatory measurement of 3D knee joint angle. J. Biomech. 1029–1035 (2008). https://doi.org/10.1016/j.jbiomech.2007.12.003
14. Lee, W.H., Lee, R.B.: Implicit smartphone user authentication with sensors and contextual machine learning. In: International Conference on Dependable Systems and Networks (DSN) (2017). https://doi.org/10.48550/arXiv.1708.09754

15. Buriro, A., Crispo, B., Zhauniarovich, Y.: Please hold on: Unobtrusive user authentication using smartphone's built-in sensors. In: International Conference on Identity, Security and Behavior Analysis (ISBA), Delhi, India (2019). ISBN: 978–1–5090–5592–0, https://doi.org/10.1109/ISBA.2017.7947684
16. Taylor, K., Silver, L.: Smartphone Ownership Is Growing Rapidly Around the World, but Not Always Equally, Pew-Research Center (2019)

Comparative Analysis of Machine Learning Algorithms for COVID-19 Detection and Prediction

Shiva Sai Pavan Inja, Koppala Somendra Sahil, Shanmuk Srinivas Amiripalli$^{(\boxtimes)}$, Viswa Ajay Reddy, and Surya Rongala

Department of Computer Science and Engineering, GST, GITAM University, Visakhapatnam, AP, India

{121910307009,121910307036,121910307055}@gitam.in, shanmuk39@gmail.com

Abstract. The outbreak of the Covid-19 happened in the year 2020 and the absence of treatment of the virus has motivated researchers to make a thorough study of the virus and its transmission. Researchers belonging to the field of Machine Learning have started implementing various algorithms to be able to identify or anticipate if the virus will be present in the human body. Data Science and Machine Learning are the leading fields of Computer a that would help in the process of detection and prediction of occurrence of events. Our paper is a comparative study of algorithms for COVID-19 analysis. The various algorithms that we have chosen are Logistic Regression, Random Forest, Decision tree and Support Vector Machine. In this study, we developed an interface that would allow the user to select the method they wish to apply to the Covid-19 dataset. As per the requirements of the user, he would be selecting and declaring the most efficient algorithm based on accuracy, precision, F1 score and recall. Based on the symptoms that the user selects on the interface, our model would also be able to forecast whether covid-19 is present in the human body.

Keywords: Logistic Regression,COVID 19 · Support Vector Machine · Decision tree · Random Forest

1 Introduction

The outbreak of the pandemic Covid-19 in the year 2020 has created havoc across the globe. Various researchers belonging to the field of Machine Learning have started their investigation on SARS-CoV-19 Virus to understand how the algorithms would help them in predicting the virus as it became an epidemic with no treatment initially. The fatal virus was spreading at a very rapid pace and infected millions of people across the globe. As per the data provided by World Health Organization (WHO), as of the first day of February 2023, there have been 6,813,845 fatalities and 753,651,712 verified cases of COVID 19 [1, 2]. There was also a decline in the economy of all the countries as a lockdown was imposed and all airlines were grounded inorder to stop the human to human transmission

P. Das et al. (Eds.): AMRIT 2023, CCIS 1954, pp. 147–156, 2024.
https://doi.org/10.1007/978-3-031-47221-3_14

of the virus. Post the outbreak of the virus, researchers have started studying various parameters that would make a human body suitable to get infected. Various researchers from countries like India, China, Turkey, USA etc. [3]. Have published articles on Covid-19. Few studies have argued that the geographical location of a country can affect the virus. According to Dr. Aaron Berstein, the director of Harvard Chan C-CHANGE, there is no proof that viruses are spreading more rapidly due to climate change. Although there are various animals that migrate to different regions with changes in seasons and through the migration the animals bring various pathogens with them that might change their hosts as they come to a new region. A branch of artificial intelligence called "machine learning" focuses on applying algorithms to empirical data, allowing computers to create models for complex relationships or patterns without having to be explicitly programmed [4]. It is founded on the idea that behavior may be replicated and fed by a lot of data. The conventional machine learning procedure required a train dataset, test dataset, and algorithm in order to create a model for making predictions. Machine learning has two subcategories: supervised learning and unsupervised learning. The goal of supervised learning is to continuously close the gap between the model's predicted output and actual outputs by training the model using labels. By utilising techniques like Support Vector Machine and Logistic Regression, it is used to handle problems involving both classification and regression. The result is referred to as Classification if the labels take the form of distinct classes. If the labels are in continuous quantity, the output is called Regression [5]. In Covid-19 detection, there are various symptoms that are used as parameters in our paper such as breathing problems, chronic lung disease, heart disease, diabetes, hypertension etc. [6]. As mentioned earlier, there various algorithms that were used in our paper for covid-19 analysis are Logistic Regression, Support Vector Machine algorithms that belong to classification and Decision Tree and Random Forest algorithms that belong to Regression problem [7].

2 Literature Review

In this paper i.e. comparative study of algorithms for covid 19 analysis has not been an easy task and a lot of research was conducted to get deep insights regarding the algorithms that we were trying to implement. We went through various research papers that were published over a particular algorithm by various authors in the past few years. Our study of those algorithms through those papers has been a great help in this paper as we understood thoroughly regarding the implementation of each algorithm and trained all the models to achieve the objective of the paper [8]. Finally, we have successfully created the interface that allows the user to choose the algorithms or other parameters through which the user would declare the best algorithm. Furthermore, our model would also predict the presence of virus in the human body based on the symptoms selected by the user. The table below contains all the information related to the research papers that were used by us to finish our paper [8]. Creating and training models on the dataset selected is definitely a tough row to hoe but these papers have guided us and helped us in solving the conflicts that have arisen while doing this paper [10].

The field of machine learning has seen a surge of interest in the past couple of years due to the COVID-19 pandemic. Many researchers have attempted to predict the spread of the virus using various algorithms and techniques. In this literature survey, we review eight recent research studies that have used machine learning algorithms to predict COVID-19. In this paper authers S. Albert Antony Raj and L. William Mary in 2021, compared Naive Bayes, Machine, K-Nearest Neighbors, and Support Vector algorithms for predicting COVID-19 [11]. The study declared the best algorithm for prediction based on F1 score, precision, accuracy, and recall. However, the study ignored other important algorithms for prediction. In this paper authers Arpita Chakraborty, Soham Guhathakurta, Souvik Kundu, and Jyoti Sekhar Banerjee in 2021, briefly focused on the Support Vector Machine algorithm for predicting COVID-19 [12]. Although the study provided results for precision, recall, F1-score, and support, it did not explore the potential of other algorithms.In this paper authers Vishan Kumar Gupta, Dinesh Kumar, Avdhesh Gupta, and Anjali Sardana in 2021, compared five algorithms, namely, Multinomial logistic regression, Random Forest, SVM, Decision tree, and neural networks with R programming to predict COVID-19 [13]. The study only considered accuracy and ignored other parameters. In this paper authers Narayana Darapaneni, Vaibhav Kherde, Deepali Nikam, Swanand Katdare, Kameswara Rao, Anagha Lomate, Anwesh Reddy Paduri, and Anima Shukla in 2021, used the Seird model, Logistic Regression, and ARIMA model to predict COVID-19 using data published by the Covid-19 India API. The study used R-squared and Root Mean Square Error to indicate how well the regression model fits the data. However, the sensitivity of the Root Mean Square Error to outliers was not considered.In this paper authers Anika Bhardwaj and N. Natarajan in 2021, compared Gaussian Naive Bayes, Decision Tree, Logistic Regression, Multilinear Regression, SVM, KNN + NCA, XGB classifier, and Random Forest Classifier to predict COVID-19 in India [14]. The study considered accuracy, precision, and coefficient of determination and concluded that Random Forest and Random Forest Classifier surpass other models. However, the study compared too many algorithms and could not focus on one specific method.TIn this paper authers Amir Ahmed, Ourooj safi, Entisar Alkayal, Sharaf Malebary, and Sami Alesawi in 2021, focused on the Decision Tree algorithm to predict COVID-19 using laboratory findings of Hospital Israelita Albert Einstein in São Paulo, Brazil. The study compared various decision tree ensembles using accuracy, precision, recall, and F1-Score. However, too many parameters were compared for many decision tree ensemblers.In this paper authers V Vaishnavi, B. Vinay Kumar, S. Samyukta, V Mounika Ravali, and P. Sarayu in 2022, compared Linear Regression, Lasso Regression, Exponential Smoothing, Single, double, and triple exponential smoothing to predict the outbreak of the virus [15]. The dataset used in this study was obtained from the John Hopkin universityGithub Repository. The study provided various measures, such as Mean absolute error, R squared score, Root mean square error, and Mean square error, which could help authorities take necessary action to control the outbreak of the virus. However, the prediction of the outbreak [16].

3 Proposed Model

This comparative study of machine learning algorithms compares and evaluates how well various machine learning algorithms perform on predicting the covid 19 disease based on the symptoms a person has. The goal of the study is to determine each algorithm's advantages and disadvantages as well as to select the best algorithm for a certain task [17]. Such a research can be helpful in a number of ways: Algorithm Selection: The best algorithm for Covid-19 prediction may be selected by evaluating the performances of alternative algorithms. Performance Comparison: This research can offer a quantitative assessment of the effectiveness of several algorithms and can assist in determining which algorithm is the most precise, effective, and reliable for Covid-19 prediction. Model Selection: The study can shed light on which model parameters or hyperparameters work well for a specific method, which can help with model choice and optimization. Advancement of the Field: By introducing new algorithms, refining current algorithms, or merging algorithms to build hybrid models, the study's findings can be utilised to improve the field of machine learning. A comparison of machine learning algorithms may be made by using various performance indicators, including F1 score, accuracy, recall and precision [18]. The findings of a comparative research must be viewed in the context of the particular topic, data, and assessment approach employed; as a result, the findings may not generalise to other situations or datasets (Fig. 1).

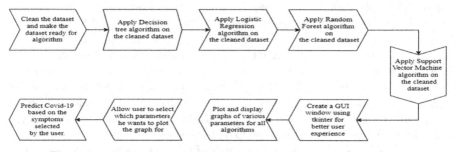

Fig. 1. Process flow of comparative study of algorithms on covid-19 dataset

3.1 Proposed Algoritham

Input: Covid-19 dataset and symptoms of user Output: Comparative analysis of machine learning algorithms and Covid-19 prediction.
Step1:Start *Step2: Clean the dataset and make the dataset ready for algorithm* *Step3: Apply Decision tree, Logistic Regression, Support Vector machine and* *Random Forest algorithm on the cleaned dataset* *Step4: Create a GUI window using tkinter for better user experience* *Step5: Allow the user to select the algorithm he wants* *Step6: Allow the user to select the parameters required* *Step7: Plot and display graphs of various parameters for all algorithms* *Step8: Allow user to select which parameters he wants to plot the graph for* *Step9: Allow the user to select the symptoms of the patient* *Step10: Predict Covid-19 based on the symptoms selected by the user.* *Step11: Stop*

In this research paper, the aim to identify and predict the presence of COVID-19 in human tissue by utilizing machine learning methods. They employ four algorithms, namely Decision Tree, Random Forest, Support Vector Machine, and Logistic Regression, and provide an interface that allows users to choose the most suitable algorithm based on their requirements and symptoms [19]. The authors illustrate the process of preparing the dataset, splitting it into training and test sets, and creating machine learning models using each of the selected algorithms. To compare the performance of the four algorithms, the authors use different performance metrics such as Accuracy, F1-score, precision, and recall. These metrics are visually presented for easy comparison. Among the four methods, the Support Vector Machine approach is found to be the most effective in terms of accuracy, while the Decision Tree algorithm is the best in terms of all four performance parameters. To demonstrate the features of all four methods and predict the presence of COVID-19, the authors also developed a graphical user interface (GUI). The GUI provides a user-friendly platform to choose the most suitable algorithm based on the user's specific requirements and symptoms.Overall, this research contributes to the use of machine learning methods to detect and forecast the presence of COVID-19 in human tissue. The authors' approach provides a user-friendly and effective way to choose the most suitable algorithm for predicting the presence of COVID-19. The paper's findings can be useful for healthcare professionals and researchers working in the field of COVID-19 diagnosis and treatment [20].

4   Results and Discussion

In this study, the dataset was first cleaned and non-numeric values were converted into numeric values using Scikit Learn's LabelEncoder module. The modified dataset was then split into training and test sets, with the training set being used to train machine learning models. Four algorithms were used to create machine learning models, and their parameters were compared in Table 1. The decision tree method achieved an accuracy of 98.3%, with precision, recall, and F1-score of 99.7%, 98.3%, and 99.0%, respectively. The Random Forest algorithm achieved an accuracy of 97.9%, with precision, recall, and F1-score of 99.2%, 98.3%, and 98.8%, respectively. The accuracy, precision, recall, and F1-score of the Logistic Regression algorithm were 96.8%, 97.1%, 99.0%, and 98.1%, respectively. The SVM algorithm scored 99.1% F1-score, 99.3% Precision, 98.9% Recall, and 98.5% Accuracy. Graphs are used to give an accurate comparison of all algorithms' parameters. The paper describes the development of a graphical user interface (GUI) to predict the existence of COVID-19 in human tissue using four machine learning algorithms. Figure 2 shows the main GUI window, which includes four sub-figures: Fig. 3 displays the accuracy of the four ML models, Fig. 4 shows the precision of the four ML models, Fig. 5 presents the recall of the four ML models, and Fig. 6 shows the F1-score of the four ML models. These figures provide a visual representation of the performance of each algorithm in terms of the four performance parameters, allowing users to easily compare and choose the most suitable algorithm for their needs. In addition to the main GUI window, the paper also presents Fig. 7, which represents the window for predicting COVID-19 in a person. This window allows users to input various symptoms and medical conditions of a patient, and based on the selected machine learning algorithm, predicts the existence of COVID-19 in the person. The GUI provides a user-friendly and intuitive way to perform COVID-19 prediction, making it more accessible to healthcare professionals and the general public (Fig. 8, Fig. 9, Fig. 10, Fig. 11).

Table 1. Parameters of different Machine Learning models

Sno	Model name	Accuracy	Precision	Recall	F1-Score
1	Decision Tree	98.3	99.7	98.3	99.0
2	Random Forest	97.9	99.2	98.3	98.8
3	Logistic Regression	96.8	97.1	99.0	98.1
4	Support Vector Machine	98.5	99.3	98.9	99.1

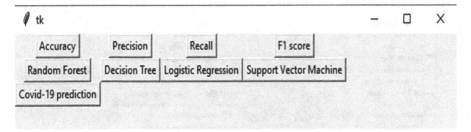

Fig. 2. Represents the main GUI window

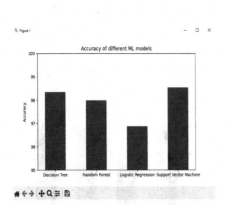

Fig. 3. Accuracy of the four ML models

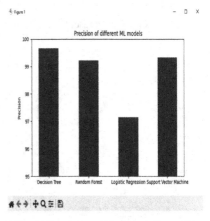

Fig. 4. Precision of the four ML models

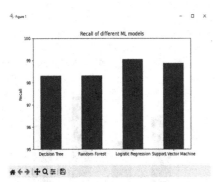

Fig. 5. Recall of the four ML models

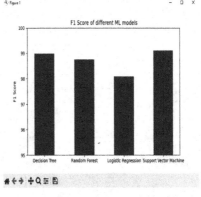

Fig. 6. F1-score of the four ML models

154   S. S. P. Inja et al.

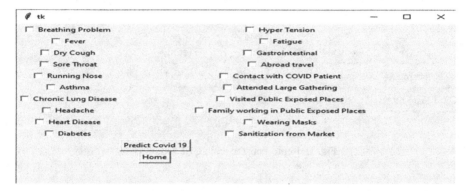

Fig. 7. Represents the window for predicting Covid-19 in a person

Fig. 8. Prediction of Covid is done based on symptoms selected

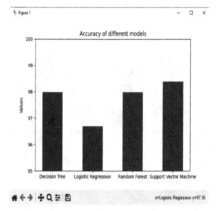

Fig. 9. The corresponding graph is plotted

Fig. 10. Prediction of Covid is done based on symptoms selected

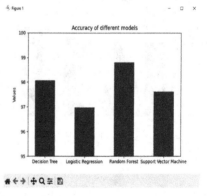

Fig. 11. The corresponding graph is plotted

5 Conclusion

The performance of four different machine learning models for the prediction of Covid-19 using a dataset of Covid-19 symptoms is compared in this study. The Decision Tree model demonstrated superior performance, achieving an accuracy of 98.3%, precision of 99.7%, recall of 98.3%, and F1-score of 99.0% on this dataset. Additionally, a graphical user interface (GUI) was developed using the Tkinter module in Python to provide the user with a visual representation of various parameters, including accuracy, precision, F1-score, and recall, for different machine learning algorithms. The GUI allows users to select specific parameters and plot graphs between the selected parameters. By choosing the symptoms that the subject is displaying, users can also forecast the existence of Covid-19. The GUI predicts the presence of Covid-19 using all four algorithms and displays their respective accuracies to assist the user in selecting a reliable algorithm.

References

1. Mary, L.W., Raj, S.A.A.: Machine learning algorithms for predicting SARS-CoV-2 (COVID-19)–a comparative analysis. In: 2021 2nd International Conference on Smart Electronics and Communication (ICOSEC), pp. 1607–1611. IEEE (2021)
2. Guhathakurata, S., Kundu, S., Chakraborty, A., Banerjee, J.S.: A novel approach to predict COVID-19 using support vector machine. In: Data Science for COVID-19, pp. 351–364. Academic Press (2021)
3. Gupta, V.K., Gupta, A., Kumar, D., Sardana, A.: Prediction of COVID-19 confirmed, death, and cured cases in India using random forest model. Big Data Mining Anal. **4**(2), 116–123 (2021)
4. Darapaneni, N., et al.: Coronavirus outburst prediction in India using SEIRD, logistic regression and ARIMA model. In: 2020 11th IEEE Annual Ubiquitous Computing, Electronics & Mobile Communication Conference (UEMCON), pp. 0649–0655. IEEE (2020)
5. Bhardwaj, A., Natarajan, N.: Covid-19 data analysis using machine learning. In: 2021 3rd International Conference on Advances in Computing, Communication Control and Networking (ICAC3N), pp. 2096–2099. IEEE (2021)
6. Sharma, S.K., Lilhore, U.K., Simaiya, S., Trivedi, N.K.: An improved random forest algorithm for predicting the COVID-19 pandemic patient health. Ann. Roman. Soc. Cell Biol. 67–75 (2021)
7. Ahmad, A., Safi, O., Malebary, S., Alesawi, S., Alkayal, E.: Decision tree ensembles to predict coronavirus disease 2019 infection: a comparative study. Complexity, 1–8 (2021)
8. Vaishnavi, V., Samyuktha, S., Ravali, V.M., Sarayu, P., Kumar, B.V.: Future prediction of COVID-19 based on supervised machine learning models. In: 2022 6th International Conference on Trends in Electronics and Informatics (ICOEI), pp. 1171–1177. IEEE (2022)
9. Prakash, S., Pathak, P., Jalal, A.S.: Predicting COVID-19 fourth wave incidence in India using machine learning algorithms and SEIR model. In: 2022 IEEE 9th Uttar Pradesh Section International Conference on Electrical, Electronics and Computer Engineering (UPCON), pp. 1–6. IEEE (2022)
10. Jaidhan, B.J., Madhuri, B.D., Pushpa, K., Devi, B.L.: Application of big data analytics and pattern recognition aggregated with random forest for detecting fraudulent credit card transactions (CCFD-BPRRF). Int. J. Recent Technol. Eng. **7**(6), 1082–1087 (2019)
11. Amiripalli, S.S., Kollu, V.V.R., Prasad, R., Jitendra, M.S.: A Mathematical-based epidemic model to prevent and control outbreak of corona virus 2019 (COVID-19). Big Data Analy. Mach. Intell. Biomed. Health Inform. Concepts Methodol. Tools Appl. 187–203 (2022)

12. Boddeda, R.L.N., Prasad, R., Amiripalli, S.S., Jitendra, M.S.: Prediction of diabetes using hybridization based machine learning algorithm. In: 2023 International Conference on Smart Computing and Application (ICSCA), pp. 1–5. IEEE (2023)
13. Mondal, M.R.H., Bharati, S., Podder, P.: Diagnosis of COVID-19 using machine learning and deep learning: a review. Curr. Med. Imaging 17(12), 1403–1418 (2021)
14. Thota, J.R., Kothuru, M., Shanmuk Srinivas, A., Jitendra, M.S.N.V.: Monitoring diabetes occurrence probability using classification technique with a UI. Int. J. Sci. Technol. Res. 9(4), 38–41 (2020)
15. Ivanov, D.: Predicting the impacts of epidemic outbreaks on global supply chains: a simulation-based analysis on the coronavirus outbreak (COVID-19/SARS-CoV-2) case. Transp. Res. Part E Logist. Transp. Rev. 136, 101922 (2020)
16. Srinivasu, P.N., Norwawi, N., Amiripalli, S.S., Deepalakshmi, P.: Secured compression for 2D medical images through the manifold and fuzzy trapezoidal correlation function. Gazi Univ. J. Sci. 1 (2021)
17. Muhammad, L.J., Algehyne, E.A., Usman, S.S., Ahmad, A., Chakraborty, C., Mohammed, I.A.: Supervised machine learning models for prediction of COVID-19 infection using epidemiology dataset. SN Comput. Sci. 2, 1–13 (2021)
18. Shaban, W.M., Rabie, A.H., Saleh, A.I., Abo-Elsoud, M.A.: A new COVID-19 patients detection strategy (CPDS) based on hybrid feature selection and enhanced KNN classifier. Knowl.-Based Syst. 205, 106270 (2020)
19. Jitendra, M.S., Srinivasu, P.N., Shanmuk Srinivas, A., Nithya, A., Kandulapati, S.K.: Crack detection on concrete images using classification techniques in machine learning. J. Crit. Rev. 7(9), 1236–1241 (2020)
20. Sungheetha, A.: COVID-19 risk minimization decision making strategy using data-driven model. J. Inf. Technol. 3(01), 57–66 (2021)

Machine Learning Classifiers Explanations with Prototype Counterfactual

Ankur Kumar$^{(\boxtimes)}$, Shivam Dwivedi$^{(\boxtimes)}$, Aditya Mehta$^{(\boxtimes)}$,
and Varun Malhotra$^{(\boxtimes)}$

Data Science, Capri Global Capital Limited, Gurugram, India
{ankur.kumar,shivam.dwivedi,aditya.mehta,varun.malhotra}@capriglobal.in
https://www.capriloans.in/

Abstract. Machine learning applications have rapidly entered our day-to-day lives, and the stakes of input from these applications have increased for various highly valued decision-making processes. Explanation of these applications has been a prime focus in recent years for better interpretability, fairness, and reliability. In this paper, using the prototype, we presented a novel approach for counterfactual generation, resulting in better proximal instances with fewer variable characteristics and interpretability metrics that enable a better understanding of the models. We have empirically presented the result of the counterfactual generator and interpretability metric on tabular and image datasets, i.e., the Adult-Income, MNIST, and Breast Cancer datasets. We compared our proposed results to KD-Tree, and the Genetic Algorithm, demonstrating improved interpretability and well-approximated counterfactuals.

Keywords: Machine Learning · Counterfactual · Explainable AI · Interpretability

1 Introduction

Humans are extraordinary creatures because they can speculate about how an event might have played out differently without experiencing it. Such counterfactual reasoning aids us in comprehending the past, devising tactics, forming opinions about people and situations, and directing executive function [7]. To address the lack of transparency and interpretability of AI applications, the human behavior of searching for alternatives has been incorporated into the interpretability of AI-based models, which has been widely proven and accepted [5,6,12]. We, as humans, consider the possibility of different outcomes based on "what if antecedents were different." There are several examples, such as: What factors would not have led to an accident? or which symptoms point to a different diagnosis? and much more involving decision-making. To answer such queries and change the altered outcome, it's essential to understand what changes are

---

S. Dwivedi and A. Mehta — equal contribution.

P. Das et al. (Eds.): AMRIT 2023, CCIS 1954, pp. 157–169, 2024.
https://doi.org/10.1007/978-3-031-47221-3_15

required; this provides better insight and interpretability, thus greater transparency over the deduction. This is widely known as the counterfactual explanation [14].

Consider an example of a person applying for a home loan for the amount of $100,000, has been rejected by the bank. What are possible feature changes that would result in a change in the resultant outcome from rejection to the approval of the loan? A counterfactual explanation [8,14] provides such an answer based on the similar patterns of customers and provides the changes that are required to be made, such as increasing the monthly salary or reducing the FOIR (fixed obligations to income ratio). These counterfactuals are not easy to generate, as the idea of counterfactuals is not to generate any far-fetched alternatives. But should consider the ease of change of features. To regulate the generated counterfactuals that solve the stated problem above, parameters such as proximity, diversity, and sparsity [13] and the feature controlling set are used. Proximity results in the closest result to the input instance; diversity maintains the heterogeneity of modifications, and sparsity limits the number of features to be modified.

In the paper, we propose a counterfactual generations optimization function by introducing a class prototype with the regularised unit and relative change in feature. The generation of closest counterfactuals is based on the characteristic that is closest to the decision boundary. We regulate the stages in our algorithm using two controlling features: the degree of shift and relative change along all the other independent feature sets. The potential energy term has been introduced to control the quantity of step energy needed to move toward the target class and the elasticity term keeps the proportional change among the independent features. The objective of the prototype is to generate samples that are closest to the target class deciding the direction of the objective function.

2   Related Work

Recent research has also explored the use of generative models for generating heterogeneous counterfactuals. It is motivated by the fact that a single counterfactual explanation does not always provide a complete understanding of the model's reasoning and multiple explanations can help to provide a more panoramic perspective. These methods involve training a generative model on the input-output pairs of the original model and then using the generative model to generate counterfactuals. This approach has the advantage of being able to generate counterfactuals that are not only different from the original input but also realistic and human-understandable. There is other counterfactual research, such as that by Delaney et al. [3], who discuss counterfactuals in the context of time series data and develop a Native Guide as a solution approach, which adapts existing counterfactual instances with details of the characteristics of good counterfactuals and discusses the superiority of the Native Guide compared to existing techniques for time series data. Dandl et al. [1] also discussed multi-objective counterfactual explanations, an approach that allows for a better trade-off between different objectives and is based on the multi-objective opti-mization literature. An adjusted version of the non-dominated sorting genetic algorithm (NSGA-II) [2] is used to solve the multi-objective problem. A further idea of counterfactual evolved to Diverse counterfactual explanations [13] where it can also be useful for identifying potential biases in a model's predictions, as well as for improving the robustness of the model Mothilal et al. [13] have high-lighted this idea. Here they have presented the idea of diverse counterfactuals and compared the result with the standard baseline methods such as LIME etc. The generation of counterfactual examples depends upon proximity, sparsity, and diversity metrics. We have taken the idea of these metrics and generated the counterfactuals with prototypes. Looveren et al. [10] introduced two inter-pretability metrics that measure the similarity between the sample space distri-bution and counterfactual examples. In the paper, they have presented results on these metrics along with several sets of objective functions introducing pro-totypes. So it can be understood that there are various methods of generation such as KD-Tree, Gradient Descent, and Genetic Algorithm for the generation of the counterfactuals.

3   Methodology

In this section, we present the optimization loss function used to generate counterfactuals. This function includes prototype loss, sparsity, and proximity. The goal is to create k sets of counterfactuals $(cf_1, cf_2, \ldots, cf_k)$ that are d-dimensional. The inputs for this problem are a query instance (x) and a machine learning model whose decision for each counterfactual is different from that of the query instance. The model is assumed to be a binary classification model throughout the paper, however, the one-vs-one or one-vs-rest methods can be applied to extend the approach to multi-class problems. To ensure that the coun-terfactuals are actionable, feasible, and closest to the original sample space, we have adapted the loss metric from Mothilal et al. [13] and combined it with the best-performing loss functions from Looveren et al. [10]. This includes a penalty algorithm and a relative change in independent variables. In the following sec-tions, we will discuss in detail our methodologies and their inspirations.

3.1   Background

To find the counterfactual instance cf_i such that both counterfactuals and query instance $c_i \in A$ where A represents the d-dim feature space. The methodology which generates the diverse counterfactuals proposed in Mothilal et al. [13] does not work well for the non-differentiable machine learning models. To overcome this problem we have introduced a prototype function to replace the hinge loss and introduced ΔC_i for capturing relative change in independent variables in Sect. 3.2

3.2 Objective Loss Function

We define the objective loss function as given below:

$$L = \underset{cf_1,\ldots,cf_m}{\arg\min} \left[\underset{proto}{\arg\min} \|c_f - F(c_f, \text{proto})\|^2 + \lambda_1 * \text{Potential Energy} \right.$$

$$\left. + \frac{\lambda_2}{d} \sum_i \Delta C_i \right] \quad (1)$$

Each term's definition and advantage are defined as follows:

$F(\mathbf{c_f}, \mathbf{proto})$ To generate the prototype of the counterfactual we follow the below stated steps:

Algorithm 1. Selection of Prototype

Require: Generation prototype for the Counterfactual Example Generation

Step-1: Select the list of k closest sample prototypes of the target class from the sample space using Euclidean Distance.

Step-2: Arrange the list in non-decreasing order.

Step-3: Pop a prototype from a sorted list and pass Optimisation Function.

Step-4: If the Counterfactual Example from the provided prototype is not generated then go to step 3.

Step-5: If Counterfactual Example generated then exit.

Step-6: If the list is empty then CF is not possible.

It is the Euclidean distance between the query instance and the prototype example. The prototype example is obtained by taking the featurewise mean of the k-nearest examples of the desired class. Thus, the prototype acts as the localized centroid of the nearest desired class cluster. As a result, it helps perturb the counterfactual towards the prototype center of the desired class. Initially, the counterfactual example is initialized with the query instance, and slowly it should move toward the prototype example.

$$c_f = \begin{cases} \text{query instance } (x), & \text{initialization} \\ c_f - \frac{\partial L}{\partial c} & , \text{ otherwise} \end{cases} \quad (2)$$

Potential Energy. We have introduced a penalty term, which can be conceptualized as potential energy, to limit the counterfactual example going far from the desired range for the feature. Since each feature has different difficulties in changes, we allow our objective function to take this into consideration. In our experiments, we have taken a constant change It is calculated as defined in Algorithm 2

Algorithm 2. Definition of Potential energy for generated counterfactual

---

Require: Percentage change allowed each feature [ch_1, ch_2 ...] and Counterfactual example

Step-1: Calculate the unit value of feature i as follows:

$$unit_i = (max_i - min_i)/100$$

Here min_i and max_i are minimum and maximum values for feature i respectively
Step-2: calculate the min and max acceptable value for feature i based on the change allowed calculated as below:

$$min_i = x_i - unit_i * ch_i \%$$

$$max_i = x_i + unit_i * ch_i \%$$

Step-3: Now, for each feature, calculate the inertia as:

$$\text{Potential Energy} = \begin{cases} 0 \text{ if } cf_i \in [min_i, max_i] \\ |cf_i - max_i| \text{ if } cf_i > max_i \\ |min_i - cf_i| \text{ otherwise} \end{cases}$$

Step-4: Calculate L_2 norm of Potential energies.

ΔC_{f_i}: **Elasticity/Relative Change in Independent Features.** In most of the counterfactual literature, the focus is on the proximity and feasibility of individual features in isolation. This, however, leads to the generation of improbable counterfactual examples. To address this issue, we introduce the term **elasticity** in the objective function: the change in correlation values among the features can be controlled. Let's say we have a data set with these features: $x_1, x_2, x_3, \ldots, x_d$ We start with creating a partial dependence regression for each of the individual features. We can write this dependence regression for i^{th} feature as below:

$$X = [x_1, x_2, \ldots, x_{i-1}, x_{i+1}, \ldots, x_D, 1] \tag{3}$$

$$W = [w_1, w_2, \ldots, w_{i-1}, w_{i+1}, \ldots, w_D, b]^T \tag{4}$$

$$x_i = X.W \tag{5}$$

Now once we have a counterfactual c_f. We can assume the counterfactual would be able to retain the partial dependence as the original dataset substituting the value of c_f in the Eq. 3 we get:

$$C_f = [x_1, x_2, \ldots, x_{i-1}, x_{i+1}, \ldots, x_D, 1] \tag{6}$$

$$W = [w_{i1}, w_{i2}, \ldots, w_{ii-1}, w_{ii+1}, \ldots, w_D, b_i]^T \,; \tag{7}$$

$$\tilde{c}_i = C_f.W \tag{8}$$

We define ΔC_{f_i} as the difference between value obtained in Eq. 9 and the actual counterfactual value obtained for feature i

$$\Delta C_{f_i} = |\tilde{c}_i - cf_i| \tag{9}$$

4   Evaluating Counterfactuals

It is important to assess the interpretability, sparsity, and feasibility of counterfactual explanations. We use the interpretability metrics IM1 and IM2 described by Looveren at el. [10] to assess interpretability. We use a novel iterative approach to vary the feature in pairs, so the sparsity is always maintained by our approach. For feasibility, it is important to make sure that the features that are perturbed maintain a correlation with the other features. If this correlation is not maintained, then the counterfactuals that are generated might not be feasible. We have introduced two evaluation matrices (CEM1 and CEM2) to evaluate the counterfactuals in this aspect.

4.1   Validity, Proximity, Sparsity, and Diversity

Validity, Proximity, Sparsity, and Diversity are very commonly used to assess the quality of the counterfactuals. Validity is the ratio of the total number of CF created to the total number of query instances. Proximity is the distance between the query instance and the produced CF. Sparsity is an inverse fraction of the total number of changing characteristics. Diversity is the average difference between the produced CF.

4.2   IM1 and IM2 [10]

IM1 defines the ratio of the difference between the CF and predicted class distribution with the difference of CF distribution from the original class and in IM2 we find the distance between the CF and the overall sample space. For this we use, two types of autoencoders were used: a) classwise trained autoencoders and b) overall data-trained autoencoders. The percentage of the difference between the instance and class autoencoders to the total trained autoencoder is provided by IM1. The difference in reconstruction between overall and classwise reconstruction to the query instance is represented by IM2. The equations are 10,11

$$\text{IM1}(x_{cf}, y, y') = \frac{\|x_{cf} - \text{AE}_{y'}(x_{cf})\|_2^2}{\|x_{cf} - \text{AE}_y(x_{cf})\|_2^2 + \epsilon} \tag{10}$$

$$\text{IM2}\,(x_{cf}, y, y') = \frac{\|\text{AE}_y\,(x_{cf}) - \text{AE}\,(x_{cf})\|_2^2}{\|x_{cf}\|_1 + \epsilon} \tag{11}$$

4.3   CEM1

We define this evaluation metric so that we are able to capture the relationship between the features in the generated counterfactuals. We start with rewriting the partial dependence regression with assuming no bias for each feature as described in Eq. 12

$$x_i = w_{i1}x_1 + w_{i2}x_2 + \cdots + w_{i\,i-1}x_{i-1} + w_{i\,i+1}x_{i+1} + \ldots \tag{12}$$

For the d dimensional counterfactual example c, We can get an estimate of weights of regression as below:

$$w'_{ij} = \frac{c_{f_i}}{c_{f_j}} \tag{13}$$

This can be visualized as if while calculating the w'_{ij} only i^{th} and j^{th} feature have a dependence on each other. This w'_{ij} should be close to w_{ij}. We define the score as follows:

$$\text{score} = \frac{1}{d}\sum_i \frac{\sum_{j\neq i}^d |w_{ij} - w'_{ij}|}{\sum_{j\neq i}^d \max(w_{ij}, w'_{ij})} \tag{14}$$

4.4   CEM2

CEM2 is another correlation estimation matrix evaluating the similarity between the latent structures in the data across the modalities that counterfactual features are similarly correlated to the training data. In this method, Spearman's Rank Correlation [11] and Cosine distance is used for empirical results. Let the Upper triangular correlation matrix of Counterfactual and Training samples U_cf, U_t respectively are used with Rank correlation defined as **R** we have Eqs. 15, 16

$$r_s = P_{R(U_cf), R(U_t)} \tag{15}$$

$$\cos\theta = \frac{v_g \cdot v_0}{\|v_y\| \cdot \|v_t\|} \tag{16}$$

5   Dataset Description

The experiments are conducted on MNIST [4] image data set and two tabular data set namely Adults income [9] and Breast Cancer data set [15]. The detail of the datasets is as follows.

Adults Income: This dataset [9] is based on 1994 Census data. It includes features such as hours per week worked, education level, occupation, work class, race, age, marital status, and sex. The objective of the ML model is to determine if an individual's income exceeds $50,000.

Breast Cancer Wisconsin: The dataset [15] contains features that are computed from images of fine needle aspirate (FNA) of a breast mass. It contains 357 benign, 212 malignant class sample data.

MNIST: The MNIST Handwritten Dataset [4] is a collection of images of handwritten digits (0–9) used in machine learning. Each image is 28×28 pixels with a labeled digit. Being a very simple image data set, it helps us visualize the counterfactuals.

Table 1. Model Accuracy and Feature Details

Dataset	Model-Accuracy	Num cont	Num cat
Adult-Income	0.89	6	7
Breast Cancer Wisconsin	0.96	24	0
MNIST	0.97	784	0

Table 1 shows the summary of the model accuracy and the feature set description on the datasets. We have reported the accuracy metric for each dataset in the table. These models were trained on all the continuous and categorical features. We have used tree-based ensemble methods (non-linear) with hyperparameter tuning to obtain these results. These same models were later used in optimization to obtain the CF examples (Figs 1,2 and 3).

Query:

	age	workclass	fnlwgt	education	marital-status	occupation	relationship	race	sex	capital-gain	capital-loss	hours-per-week	native-country
0	65.00	3.00	170012.00	4.00	2.00	0.00	0.00	4.00	1.00	0.00	0.00	34.00	38.00

CFs-Generated:

	age	workclass	fnlwgt	education	marital-status	occupation	relationship	race	sex	capital-gain	capital-loss	hours-per-week	native-country
1	65.00	3.00	170012.00	4.00	2.00	0.00	0.00	4.00	1.00	0.00	0.00	48.06	38.00
2	65.00	3.00	266632.74	4.00	2.00	0.00	0.00	4.00	1.00	0.00	0.00	34.00	38.00

Fig. 1. Examples of generated CF in Adult Income Dataset showcasing two CF examples with change in two features.

Query:

	concave points_mean	concave points_worst	perimeter_worst	texture_worst	...	radius_mean	compactness_mean	perimeter_se	smoothness_se
1	0.04	0.10	96.05	24.64	...	12.47	0.11	2.50	0.01

CFs-Generated:

	concave points_mean	concave points_worst	perimeter_worst	texture_worst	...	radius_mean	compactness_mean	perimeter_se	smoothness_se
4	0.06	0.14	96.05	24.64	...	12.47	0.11	2.50	0.01
6	0.08	0.14	89.61	32.04	...	12.40	0.13	2.20	0.01

Fig. 2. Examples of generated CF in Breast Cancer Dataset showcasing two CF two and three feature change in respective CF.

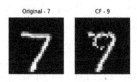

Fig. 3. Examples of generated CF in MNIST Dataset from original class of 7 to target class 9

6 Result

In this section, we present counterfactual generation from our proposed algorithm, performing a quantitative evaluation and comparison with another previously used algorithm KD-Tree, Genetic, and Gradient. To explain the model we evaluated 200 query instances quantifying on Validity, Proximity, Sparsity, Diversity, IM1, IM2, CEM1, and CEM2. In this, we have regulated the percentage change in feature values by 0.10 as mentioned in Algorithm.

6.1 Validity

The performance of CF examples generated by KD-Tree and Genetic algorithms lags, especially for Adult Income where Genetic algorithms show an upward trend but only reach a little above 60%. The validity of all the k CF examples generated by our proposed method is 100% for all the datasets.

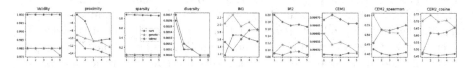

Fig. 4. Comparison of the Counterfactuals with the baseline methods on Validity, Proximity, Sparsity, Diversity, Interpretability Metrics IM1, IM2 and CEM1 and CEM2 for Breast Cancer Dataset

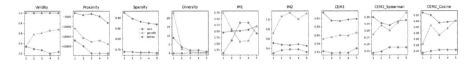

Fig. 5. Comparison of the Counterfactuals with the baseline methods on Validity, Proximity, Sparsity, Diversity, Interpretability Metrics IM1, IM2 and CEM1 and CEM2 for Adult Income Dataset

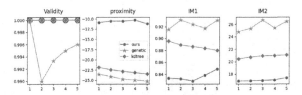

Fig. 6. Comparison of the Counterfactuals with the baseline methods on Validity, Proximity, Interpretability Metrics IM1 and IM2 for MNIST Dataset.

6.2 Proximity, Sparsity, and Diversity

As discussed in Sect. 4.1, our proposed method has better proximity with a minimal number of varying features. As per the result shown in Fig. 4 and Fig. 5, it can be observed that CF generated with our methodology has better proximity and sparsity compared to the other methodologies. The genetic algorithm shows a huge drop in valid counterfactuals as shown in Fig. 6 after the first set of examples whereas our algorithm results in better proximate with all valid counterfactuals for the MNIST dataset. As we are only concentrating on the perturbing of the continuous features in our algorithm, the diversity value obtained is less as compared to other algorithms. But the counterfactuals generated by our algorithm present a better proximate result set closer to the class boundary, leading to a better interpretation of the model, which is discussed in detail in the following section.

6.3 IM1 and IM2

In Fig. 5, the CF examples created by our technique, IM1 measure for adult income indicate that as the number of CF grows, so does the ratio of CF moving away from goal class distribution and nearer to the present class. KD-Tree and genetic, all reach around the same level after 4th CF. For IM2, it outperforms the Genetic algorithm but falls short of the KD-Tree but the CF validity of KD-Tree is very low so it doesn't add up to scoring better IM2. In Fig. 4, the CF examples for our method are close to the target class distribution but a little far from the overall distribution for one CF, but values of IM1 increased and IM2 remained higher than others, but it's important to remember that there is a trade-off between proximity, sparsity, and the interpretability metric. Figure 6, shows that the CF examples produced by MNIST are highly interpretable, outperforming the other two techniques for all K counterfactuals examples.

6.4   CEM1 and CEM2

CEM1 attempts to capture the relative change in independent features, as explained in Sect. 4.3. For our technique, the relative change is found to be low in Breast Cancer compared to others and higher in Adult Income. We have not evaluated the MNIST dataset on CEM1 and CEM2 interpretability metrics as it does not make sense to capture the correlation between image pixels. The uneven pattern is shown in Fig. 4 and Fig. 5 from our method's attempt to reduce the number of features to perturb i.e. increased sparsity and closeness. In the Breast Cancer dataset, genetic and KD-Tree sparsity approaches 0 indicating a change in all features, however, our technique generates CF with increased proximity and lower CEM scores discussed in Sect. 4.4 and Sect. 4.3 implying a shift in CF features correlation. This implies that to improve interpretability, it is necessary to examine a trade-off between supporting measures such as proximity, validity, and sparsity.

7   Conclusion

We proposed a modification to the process of generating model-agnostic counterfactual examples with the aid of class prototypes. We have introduced a cost function that combines the energy and elasticity of class prototypes, allowing for more flexible, coherent, and interpretable results. In comparison to other existing methods, our approach demonstrates notable benefits in terms of practicality, interpretability, and feasibility.

One of the key insights of our research is the trade-off between proximity and interpretability. We have provided measures to allow users to optimize for either proximity or interpretability, depending on the specific use case and desired outcome. This provides a way for the end users to trade between proximity and interpretability. To further assess the performance of our method, we have compared our results to other approaches using two novel metrics that we have introduced. These metrics aim to quantify the interpretability and feasibility of counterfactual examples, providing a more comprehensive evaluation of the effectiveness of our approach.

Our findings suggest that our proposed method outperforms existing methods on most of the metrics. This highlights the potential of our approach for providing a more interpretable and feasible solution for generating model-agnostic counterfactual examples.

8   Future Work

As we move forward, we will be working on enhancing the counterfactual generation process. The current method involves iteratively selecting which features to vary, this can be a time-consuming and credulous process. To address this issue, we plan to explore the use of Markov models or Bayesian models to streamline the feature selection process. These models can potentially help to reduce the number of iterations required and make the process more efficient.

In addition to this, while we have taken measures to capture the inter-relationship among features by adding a correlation term in our objective function, we believe that there is still room for improvement. We aim to provide a more robust and intuitive approach that is user-friendly, and that better captures the complex relationships among the features.

Our goal is to create a counterfactual generation process that is both efficient and effective. By incorporating these aforementioned techniques, we hope to make the technique more accessible to a wider range of users and to facilitate a deeper understanding of the relationships between features and their impact on outcomes. In short, our focus is to continually improve the process of counterfactual generation so that it can be used to its full potential.

References

1. Dandl, S., Molnar, C., Binder, M., Bischl, B.: Multi-objective counterfactual explanations. In: Bäck, T., et al. (eds.) PPSN 2020. LNCS, vol. 12269, pp. 448–469. Springer, Cham (2020). https://doi.org/10.1007/978-3-030-58112-1_31
2. Deb, K., Agrawal, S., Pratap, A., Meyarivan, T.: A fast elitist non-dominated sorting genetic algorithm for multi-objective optimization: NSGA-II. In: Schoenauer, M., et al. (eds.) PPSN 2000. LNCS, vol. 1917, pp. 849–858. Springer, Heidelberg (2000). https://doi.org/10.1007/3-540-45356-3_83
3. Delaney, E., Greene, D., Keane, M.T.: Instance-based counterfactual explanations for time series classification. In: Sánchez-Ruiz, A.A., Floyd, M.W. (eds.) ICCBR 2021. LNCS (LNAI), vol. 12877, pp. 32–47. Springer, Cham (2021). https://doi.org/10.1007/978-3-030-86957-1_3
4. Deng, L.: The MNIST database of handwritten digit images for machine learning research. IEEE Signal Process. Mag. **29**(6), 141–142 (2012)
5. Goebel, R., et al.: Explainable AI: the new 42? In: Holzinger, A., Kieseberg, P., Tjoa, A.M., Weippl, E. (eds.) CD-MAKE 2018. LNCS, vol. 11015, pp. 295–303. Springer, Cham (2018). https://doi.org/10.1007/978-3-319-99740-7_21
6. Guidotti, R., Monreale, A., Ruggieri, S., Turini, F., Giannotti, F., Pedreschi, D.: A survey of methods for explaining black box models. ACM Comput. Surv. **51**(5), 1–42 (2018). https://doi.org/10.1145/3236009
7. Hoeck, N.V., Watson, P.D., Barbey, A.K.: Cognitive neuroscience of human counterfactual reasoning (2015). Accessed 29 Jan 2023
8. J. Byrne, R.M.: Counterfactuals in explainable artificial intelligence (XAI): evidence from human reasoning | IJCAI. Accessed 29 Jan 2023
9. Kohavi, R., Becker, B.: UCI machine learning repository (1996). https://archive.ics.uci.edu/ml/datasets/adult
10. Looveren, A.V., Klaise, J.: Interpretable counterfactual explanations guided by prototypes. CoRR abs/1907.02584 (2019). http://arxiv.org/abs/1907.02584
11. Lovie, S., Lovie, P.: Commentary: Charles spearman and correlation: a commentary on 'the proof and measurement of association between two things'. Int. J. Epidemiol. **39**(5), 1151–1153 (2010)
12. Miller, T.: Explanation in artificial intelligence: insights from the social sciences. Artif. Intell. **267**, 1–38 (2019). https://doi.org/10.1016/j.artint.2018.07.007, https://www.sciencedirect.com/science/article/pii/S0004370218305988

13. Mothilal, R.K., Sharma, A., Tan, C.: Explaining machine learning classifiers through diverse counterfactual explanations (2019). Accessed 29 Jan 2023
14. Wachter, S., Mittelstadt, B., Russell, C.: Counterfactual explanations without opening the black box: automated decisions and the GDPR (2017). Accessed 29 Jan 2023
15. Wolberg, W., Street, W., Mangasarian, O.: Breast cancer wisconsin (1995). https://archive-beta.ics.uci.edu/dataset/17/breast+cancer+wisconsin+diagnostic

A Systematic Study of Super-Resolution Generative Adversarial Networks: Review

Ravindra Singh Kushwaha[✉] and Rajan Kakkar

School of Computer Science, Lovely Professional University, Phagwara, Punjab, India
ravindrakushwaha761@gmail.com

Abstract. Visual representation of images has the edge of analyzing very deep down in many research areas so the problems can be understood very thoroughly. Images play a key role nowadays in several scientific and non-scientific fields such as medicine, geography, astronomy, gene sequencing, climate change, global warming, etc. But every time high-quality image is not available due to some constraints such as the quality of the camera, constraints of sending and receiving images, enforcing the camera to take the pic in microseconds in the area of radiology, nuclear reactor, during volcano eruption, etc. These are some of the prestigious sectors where HR quality of image always require. For solving these problems researchers have made out a very beautiful algorithm which is an SRGAN. This method become the center of various research for enhancing the resolution of the image. Every day many people do research and publish thousands of papers on this topic. This paper utilizes those studies and done the thorough review.

Keywords: Super-Resolution · GAN · Spatial Resolution · Up sampling · Downsampling

1 Introduction

Digital imaging has been a boon and the driver of numerous technological developments. Due to advancements in picture-capturing methods, digital image processing is now suited for a wide range of applications in numerous scientific and technological sectors. Video surveillance (target recognize license plate recognition), satellite imaging, medical [34] image analysis, and computer vision are some of the main applications of digital imaging. These fields heavily rely on the images produced by image processing methods or those taken by cameras. Digital photographs, for instance, are crucial in medical applications because they help doctors diagnose patients correctly.

Similarly, it facilitates the detection of objects from the background in satellite imaging applications. The adaptability of digital images is also advantageous for computer vision applications. As time passes, There have been significant advancements in image capture methods. However, some methods are still needed in particular application areas to raise the image quality. The improvement of image quality is the main goal of this research project.

Compared to traditional photographic images, digital images utilized in medical imaging require specific methods of capture. The main differences between photography visuals and MRI scans are as follows.

P. Das et al. (Eds.): AMRIT 2023, CCIS 1954, pp. 170–186, 2024.
https://doi.org/10.1007/978-3-031-47221-3_16

1. Medical [20] photographs require meticulously controlled illumination of the subject, which is often a human, during the picture capture procedure.
2. For applications involving medical imaging, image speed is especially crucial.
3. Rather than creating aesthetically pleasant images, medical imaging helps with disease identification.
4. In medical [33] imaging, visual artifacts are far less tolerated.
5. None of these restrictions must be met for typical photographic photographs.

Why Image SR is Needed?

The basic building blocks of an image, such as its pixels or picture element, determine the quality of that image. Digital photos are divided into LR and HR categories based on their resolution. High pixel densities are present in HR photographs. The pixel density is lower in LR photos, on the other hand. As a result, high-quality photos contain more data than LR images.

In many image analysis applications, HR images are frequently needed and preferred. Therefore, these benefits result in improved human visual perception and more beneficial data for computer vision and other image analysis applications. For instance, if the object boundaries are distinct, the things of interest can be easily separated from other objects. This results in enhanced pattern recognition, image segmentation, and other performance outcomes. If HR photos are provided for medical [33] image analysis, the doctors will be able to make a precise diagnosis (Fig. 1).

Fig. 1. Shows LR imagery is present alongside an HR image [28].

As a foundation for generative models, GAN, a special kind of NN in which several networks are trained concurrently, has excelled recently [13]. The manufacturing of the SR happens from one or more blurry or noisy pics [39], and it is one of the active subfields of CSE [2]. Interpolation techniques like bicubic interpolation and Lanczos interpolation are examples of early SR techniques [14].

Zhang et al. [16] A single SR's [39] objective is to create or reconstruct a picture using HR from observation using LR. SR is the method of extracting image features from LR photographs to rebuild SR, according to authors [4]. Lin et al. [10] describe that the only function of the SR is to retrieve the lost pixels which are not attainable in classic methods.

HR photographs frequently retain more features and important information that are essential in many industries, including face identification [40], surveillance, astronomical imaging, and medical imaging [10].

According to the [10] modern state-of-the-art techniques primarily rely on supervised learning-based techniques to boost the accuracy of picture SR reconstruction. At the time when the authors have been doing research very less method are present for changing the resolution of the images.

In the computer vision field generating a SR image from the LR image become a key area in which the we are restoring the pixels. In an online setting, for example, switching from low correspondence to high correspondence can reduce the amount of storage space and network bandwidth required [7].

In real-world circumstances, the SR problem frequently has the following charac- teristics: 1) no access to HR datasets; 2) no knowledge of the downscaling approach; 3) noisy and blurry input LR photos [9]. D. Zhang et al. [7] figure out the phenomena that the we are not completely dependent for conversion of the image only one at a time we can apply the same method for transforming the image just sending multiple input and get the good quality output photos.

Utilizing two generators and discriminator pairs, CycleGAN [31, 32], has been successful in translating images from one image to another [3].

Medical images are used in conjunction with the SR [28] approach to aid in disease diagnosis [7]. The area of this technique is so wide it has touching the new milestone in everyday the authors of [14] and [20] told that the CT, MRI, ultrasound, are the focused area in medical field the high quality pixels has impacted heavily for detection, segmentation of the disease.

Biology, zoology, astronomy and astrophysics are also using these technique for getting the best qualify pixels [14].

[4, 20] Due to the low exposure of radiation during x-ray the doctors are not able to get the image perfectly so improving the visual quality of the images it is direct affecting.

Medical students mostly find the difficulty during studying the anatomical structures of the body due to poor image quality, here the [11] has done a great job to apply this technique on anatomical dataset and got the HR photographs with very good and efficient accuracy [11].

Because of physics, technological constraints, HR imaging has limitations, such as a length and duration of the time period, has restricted the spatial resolution of the photos [11, 39].

An image SR retrieve the damaged by attempting to find out the lost information. SR techniques for image processing in medicine [6] have already been adopted by several photo applications, including surveillance, face [40], iris, and object detection.

Wang et al. [12] SISR, a fundamental low-level vision problem, is gaining popularity among researchers and AI companies. SISR can generate a multiple HR pixels just one LR image (Fig. 2).

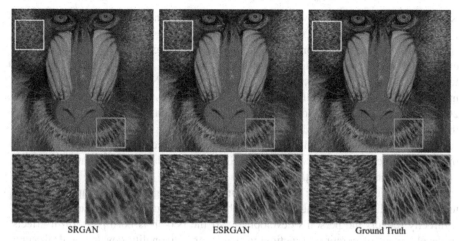

SRGAN   ESRGAN   Ground Truth

Fig. 2. Here is a comparison of different models for making the image noiseless, blurless, and highly visualizable [12, 28, 35].

2 Literature Review

Luis Salgueiro Romero et al. [17] use the WorldView-Sentinel image dataset, on that dataset they apply the ESRGAN model, they super-resolved the ESRGAN on scaling factor 5. During training of the RE-ESRGAN, the upsampling has been removed in which sensor data, satellite data is present. The evaluation of the image by the PSNR, CC value, SAM value, SSIM, the result is astounding in compare to other models. From this technique, when they analysing the visual quality of the pics, the the resolution and quality of the spatial image is high.

The MRI brain scan images are worked on by Sánchez and Vilaplana [18]. Their method is completely based on the SRGAN. The PSNR and SSIM value of 2x upsampling is 39.28 and 0.9913 and 4x upsampling of PSNR and SSIM value is 33.58, 0.988. According to their research, The PSNR value does not quality the basic criteria in compare to the SNNC got higher values, SSIM performance increase exponentially in terms of convolution resizing and upsampling, This is much similar to the human retina's visual perception. Natural image generation always a difficult part of the image but requirement forced the methods to achieve that, so satisfying this purpose they use the content loss, how much data is lost during processing, adversarial loss is about to how many time this part is not able to detect the fake samples and generator loss is to got captured by the inspector which is a adversarial neurons.

An ESRGAN has been proposed by Rakotonirina and Andry [19]. DIV2K is the training set being employed. It is a collection of images with a 2K resolution that is appropriate for SR. The DIV2K dataset initially only included 800 images. Authors used the data augmentation method for producing more synthesized data so the sample size of the dataset can also increased. Adam is accustomed to using $1 = 0.9$ and $2 = 0.999$ while optimizing the model. The training of the +nESRGAN is performed on the generator which has inbuilt 23 block with the help of pytorch and NVIDIA Tesla

K80GPUS. Their results shows that the visual perception of the output is higher on PIRM dataset.

Recently developed PyTorch 0.4.1 is used by the Bing et al. [20], in combination of the CUDA8, 5.1 and Nvidia's best processor which is 1080Ti is performed on the Ubuntu 16.04 is used for creating the recommended GAN-based medical picture SR model. Bing et al. used the DRIVE [24] dataset which has training and testing images 20 and 20 respectively. Author also uses the one more dataset STARE [25] in which there are 397 images are found, but the selection of the images for training and testing is selected randomly categorize in part A and B. The presented method beated the existing models in for getting the high accuracy in PSNR and SSIM values. The dataset of retina is also well utilized by the authors.

Delannoy et al. [21] work on neonatal brain MRI images, and they apply SRGAN techniques to that dataset. The dataset which is used is crucial in his research because it directly reflects the person's development. If the person is born pre-term it affects in development of cognitive functioning, behavior, personality development, thought process, and memorization. The author's research has done a great job, they analyzed the MRI images of the brains based on those images, and when they compare the author easily figures out that the person is born pre-term or through normal delivery.

Qian Huang et al. [22] uses the shortcut method in which the input layer is directly connected to the output layer by doing this they got the image with the good metric value as well as high spectral value. For the training, Deep CNN became the priority so that the gradient features of the image which is LR and HR can also able to help the discriminator. The experimental results show that the suggested method out perform in spatial resolution and the spectral fidelity of the image. The CASI sensor which is used for taking the sensor image data is generated the hyperspectral image dataset, on that dataset the author perform the experiment. The range of spectral resolution of the image is varying according to the requirement of the image.

In the GAN framework Kalpesh et al. [23] proposed the new solution for producing the HR image from the LR image, in his proposed work they uses the unsupervised [32] perceptual function for evaluating the quality of the image, with the help of MOS, they try to provide the realistic natural images. The entire operation is performed on the NTIRE-2020 benchmark dataset, the dataset is divided in to two sets training which is a combination of the Track-1 and Track-2 dataset and validation data which has only Track-1 collection of the data. For developing the unbiased model, they generalize the network in such a way that the real-world images and the artificially generated image are both are used during training of the model. The new develop model tested on the NTIRE2020 dataset, the accuracy they get from here quite interesting and relevant for the research. For an SR technique with unsupervised learning, they propose a typical GAN framework [32]. By introducing the MOS, it is clear that the intention of the author is that to enhancing the resolution of the image. By analysing the review of the image, they got the good response from the respondent. Their new technological model shows that the generalization of the model can work on different benchmark dataset, the purpose of the generalization is to generate the HR image which can work on every LR image dataset.

Zhao et al. [24] has work on how to achieve the color-depth SR images, at the end of their finding they got the relevant results. Generating the real world images is always challenging task, for achieving this task they regularize the generator and discriminator so that it can produce the natural images. Additionally, they solve the problems like edge detection and the smoothing problem of the image.

Kabiraj et al. [25] work on the identification and recognition of number plates. The main target of the author is to identify the digits on the number plate, it is quite difficult to recognize those number digits from the LR image. After that, the author tries to upscale the image. For fulfilling this purpose they use the ESRGAN network with a modified dense network [35] and residual [33] network for performing the experimenting on the OCR dataset. When author uses the OCR paradigm, accuracy ranges from 4% to 7% for low resolution and 84% on average for high resolution. The plate pictures' improved resolution improves the model's accuracy. The OCR model demonstrates that Pro-SRGAN and SRGAN are less effective at enhancing LR images than ESRGAN. The accuracy of digit/alphabet detection is significantly increased when an LR image of the number plate is transformed into an HR image using ESRGAN.

Mahapatra et al. [1] have done a study in the domain of the Fundus image [36] dataset. In their tests, they performed experiment on the upscaling factor of 16, and shows that the HR image can be produced from the LR on this upscaling factor. This improves the accuracy of automatic picture interpretation, particularly for small or unclear disorders and landmarks. With each image having a different set of image dimensions and 100 additional rotation and translation passes, they test their method on 5000 retinal fundus images from various sources. The dark borders on the images have been removed, and they have been resized to 1024 × 1024 pixels.

3 Methodology

3.1 Network Architecture and Design

In the SRGAN [28] design, there are mainly two networks both are competent to each other at the real time. Both components are opposite of each other. Generator is trying to produce the synthetic data in which the real as well as the fake sample both are present and try to bend the rules of the discriminator. But the discriminator's task is like a checker and checks the data which was generated by the generator is valid or fake. If the data is valid then it allows passing otherwise if it found out that the data is fake it sends it back to the generator. Through this backpropagation, again generator optimizes that data and tries to make it as if it is real data and sends it back to the discriminator. Through this entire cycle, both components of the SRGAN [30] are trained in hundreds and thousands of epochs (Fig. 3).

Generator Architecture
The network of this architecture is made of a residual NN [33]. Unlike the deeper network, where the number of layers is increased, the error rate is also increased we can able to train the neurons deeper. But the residual network is using the way to train the neurons more intensely, for fulfilling this task skip connection plays a key role in training [26] the neurons in deeper. The skip connection skips those layers which tends

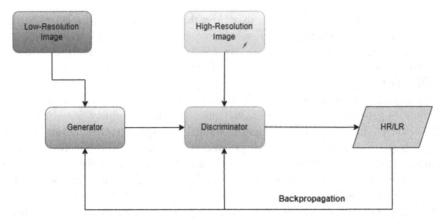

Fig. 3. Architecture of SRGAN

to produce an error during training. In Fig. 4(a.) we are showing the network design which is made of 16 residual layers, 3×3 kernel size, 64 feature maps [36], and striding is 1 for each CNN layer [40]. The batch normalization and PReLU activation functions adhere to these layers [26].

Discriminator Architecture
The aim of this is to figure out the differences between HR [33] and SR images [37]. The HR images are provided by the user during the training time and SR is inserted which was produced by the generator. It has 3×3 filter kernels on 8 convolutional layers. The factor of the kernel starts from 2 to 64 which can be increased up to 128. At the last, the output of these features is followed by the two dense layers [35], leaky ReLU [40], and sigmoid function. The sigmoid function provides the probability of the sample classification [33] (Fig. 5).

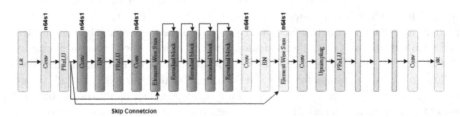

Fig. 4. Generator Architecture

A discriminator network D_{θ_D} that we alternately optimize using G_{θ_G} to resolve the adversarial min-max problem [5]:

$$\min_{\theta} \max_{\theta} E_{I^{HR} \sim p_{train}(I^{HR})} \left[\log D_{\theta_D}\left(I^{HR}\right) \right] E_{I^{LR} \sim p_G(I^{LR})} \left[\log\left(1 - D_{\theta_D}\left(G_{\theta_G}\left(I^{LR}\right)\right)\right) \right] \tag{1}$$

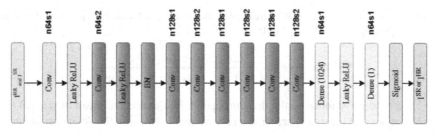

Fig. 5. Discriminator Architecture

The above equation [27, 33] is the mathematical representation of the generator and discriminator, where generator G produce the artificial as well as the real samples and discriminator D is like a judge which sample is fake or which sample is real and results are back propagated towards the generator and discriminator. By this strategy, the generator enhance its capability of generating the more realistic samples and discriminator become much more expert invigilator for addressing the issue of the fake and real sample of the data.

3.2 Evaluation Metrics

PSNR
It is one of the often employed methods to evaluate an image's quality [37] reconstruction [3, 17, 31]. The below equation define the relationship of the noisy image and the real image. In the nominator is shows the real image and the denominator representing the noisy, blurry and fake image.

$$psnr(x, \ y) = 20 \log_{10}\left(\frac{\max(x)}{\text{RMSE}(x, \ y)}\right) \tag{2}$$

Here x is described as the true image and y is defined as the estimated image.

MSE
This loss function is used to develop an ERM for the SISR. The mean squared difference between two images, expressed in pixels, is calculated as an MSE loss [3, 15]. It is seen that MSE loss. It possess the convex function property that's why some times it become very worse when the dataset is multimodal and nontrivial. Additionally, PSNR, another criterion used to analyze images, is easily increased by MSE as an object function [7].

$$\text{RMSE}(\hat{x}, \ x) = \sqrt{\frac{1}{N}\sum_{n}(\hat{x}_n - x_n)^2} \tag{3}$$

Here estimated value is $\widehat{x_n}$ and x_n is the actual value.

SSIM
It is a way to measure two photographs by taking three factors into account: brightness,

contrast, and structure [17]. The SSIM [40], a widely used index of perceived image quality, is described as follows [3, 35]:

$$SSIM(\hat{x} - x) = \frac{(2\mu_x\mu_{\hat{x}} + c_1)(\sigma_{x\hat{x}} + c_2)}{\left(\mu_x^2 + \mu_{\hat{x}}^2 + c_1\right)\left(\sigma_x^2 + \sigma_{\hat{x}}^2 + c_2\right)} \tag{4}$$

c2, c1 are used for stabilizing the division operation, the x is the mean of μ_x and the variance of the two image [3] are shown as $\sigma_x\hat{x}$, and σ_x showing the deviation of the x, [3].

MOS

This metric is used for measuring the image's class whether it is in a good class, a bad class, or just is normal class. Which was produced by the SRGAN. The methodology said that the image is shown to various people, and by looking at the image they judge the quality of the image by rating it between 1 to 5. Where 1 is the very poor quality image and 5 is the very good quality image.

Table 1. MOS Rating

Rating	Class
1	Very Poor
2	Poor
3	Fair
4	Good
5	Very Good

MOS is calculated in mathematics by averaging all single ratings given to a stimulus by test participants who participated in a subjective quality assessment examination. Thus:

$$MOS = \frac{\sum_{n=1}^{N} P_n}{D} \tag{5}$$

Here the number of people is represented by D and the particular person's rating is indicated by the P.

Zhang et al. assess the efficiency of image SR using the MOS. Given that it relies on observers' assessments of the images' quality, MOS is the most precise subjective evaluation of image standards [7].

3.3 Loss Functions

For saving the model from the local optimum, we train generator first and at the same time freeze the discriminator, and when the discriminator is trained, the generator got freezed, this entire process got repeated multiple times on various inputs and on every iteration [4]. Adversarial loss is the common function in both of them. Content loss is slowdown the process of the convergence which will affect affect the precise production of the data points [2].

For the generator training process, adversarial loss is used as the sum the weighted with content loss [4].

The loss function is the source of both adversarial loss, also known as generative loss, and content loss. The equation is provided below [27].

$$l^{SR} = \alpha l^{SR}_{Adv.} + l^{SR}_{Cont} \tag{6}$$

l^{SR}_{Cont} = Content loss
l^{SR}_{Adv} = Adversarial loss
α = Balancing loss on both sides.

Perceptual Loss
This function is used to measure the similarity between two images also measure the loss of the perceptual quality. It is a combination of the content loss function and adversarial loss function. Adversarial function is trained in such a way to distinguish in between the realistic photo and the super-resolved photos [33].

$$l^{SR} = \underbrace{\underbrace{l^{SR}_{X}}_{Content\ loss} + \underbrace{10^{-3} l^{SR}_{Gen}}_{adversarial\ loss}}_{perceptual\ loss\ (for\ VGG\ based\ content\ losses)} \tag{7}$$

adversarial loss is l^{SR}_{Gen} and content loss is defined as l^{SR}_{X}.

Adversarial Loss
Adversarial loss seeks to assist the generator as closely as feasible in bringing the SR image to the realistic HR image to deceive the discriminator [4]. During the training period, the generative loss l^{SR}_{Gen} [1] is defined as:

$$l^{SR}_{Gen} = \sum_{n=1}^{N} - \log D_{\theta_D}\left(G_{\theta_G}\left(I^{LR}\right)\right) \tag{8}$$

$\left(G_{\theta_G}\left(I^{LR}\right)\right)$ defined that the image is natural, $\log D_{\theta_D}\left(G_{\theta_G}\left(I^{LR}\right)\right)$ is the probability that the image can be natural or fake.

Content Loss
Smooth SR images produced by MSE lack high-frequency content and are perceptually unpleasant [1]. This aspect of loss is mostly in charge of an image's content restoration. For setting the content loss, we use the MSE(L1) loss for boosting the PSNR value [4]:

$$L1 = \frac{1}{N^2} \sum_{i=1}^{N} \sum_{j=1}^{N} |HR(i,j) - SR(i,j)| \tag{9}$$

The equation shows N denotes the size of the photos, assuming they are all N by N in size (Table 3).

Table 2. Comparative Analysis of SRGAN

Paper or Citation	Problem	Research Objective or Aim	Solution/Contribution	Data Collection/Dataset	Findings	Keywords
Mahapatra et al. [1]	Due to cost functions, scaling factors greater than 4 are less effective with retinal pictures in SRGAN	Transform LR retinal image into HR by upscaling factors 4, 8, 16	1. Utilizing curvature maps to construct local saliency maps [36] 2. Entropy filtering	Retinal Fundus image dataset with 5000 images, pixel size = 1024 × 1024	For disease, small, or obscured images, this method gives the precise output image	Local Saliency Maps, Retinal Image Analysis, SRGAN, ISR
Demiray et al. [2]	Due to the high cost, DEM technology is not available everywhere	To increase the resolution of the DEM images	4 times increment in spatial resolution	DEM dataset	The SISR method performs very well on the DEM dataset without requiring any additional information	DEM, DEM reconstruction, GAN, SRGAN
Tzu-An Song et al. [3]	PET produces LR, spatial, degraded, noisy, accurate image	Transform PET image into HR image	Propose SSSR for PET images	Clinical neuroimaging dataset	The SSSR method opens the gate of MRI, small scale, noisy images can be converted into SR form	SSSR, PET, Neuroimaging, Medical images, CNN
Gu et al. [4]	CT produces low-dose images in limited radiation exposure	Develop a generalized model which can work on similar types of images	Proposed a MedSRGAN method that can work on CT, and low MRI images	110 brain MRI scans and 242 thoracic CT scans are used from LUNA16, SISR dataset (Set5, Set14, and Urban100)	1. Most likely, tightly connected convolutional blocks are not appropriate for CT SR 2. Compared to RCAN, MedSR-GAN and MedSR-GAN + can achieve greater PSNR and SSIM	Brain MRI, CT scan, GAN, Low radiation exposure

(continued)

Table 2. (*continued*)

Paper or Citation	Problem	Research Objective or Aim	Solution/Contribution	Data Collection/Dataset	Findings	Keywords
Ledig et al. [5]	At high upscaling factors, how can the finer texture details of super-resolve images be recovered	To produce a 4x upscaling factor of images	First-time Present SRGAN method, Perceptual loss function	Public benchmark dataset: Set5, Set14 and BSD100	SRGAN significantly exceeds all benchmark techniques and establishes a new standard for photorealistic image processing. SR	Natural image, 4x Upscaling factor, Photo realistic texture, DRN
Ren et al. [6]	The quality of the image is degraded by many factors such as blur kernel, degradation, down-sampling method, etc	Develop the classic method of SR	1. Implemented an ensemble GAN 2. Weak supervised SR i.e. it produces HR images	NTIRE 2020	1. Good quality perceptual 2. Generating high-quality images in comparison to standard ESRGAN	Weak Supervised SR, Image Degradation, Noisy image
Zhang et al. [7]	By using MSE, the PSNR value gets high but it is not depicted the actual quality of the image i.e. blurry result and too much smoothness	For measuring the quality Perceptual loss is used instead of MSE	Propose residual-based GAN, Adversarial loss, and Convolution loss, as perceptual loss	CelebFaces [40] Attributes dataset (CelebA)	The GAN-based design reduces the disparity between the real picture and reconstructed images and creates perceptually more believable SR images	Visual quality, Realistic images,
Mahapatra et al. [8]	Small and microanatomical (retinal images) are not able to be analyzable due to LR	Producing the HR anatomical images including MRI, Retinal, etc.	It can build the image of any upscaling factor	EYEPACS, Reinal fundus images (5000), 1024 × 1024 pixel, Sunnybrook Cardiac Dataset	The experimental results show that it outperforms in comparison to others	Anatomical structure, medical images, MRI images

(*continued*)

Table 2. (*continued*)

Paper or Citation	Problem	Research Objective or Aim	Solution/Contribution	Data Collection/Dataset	Findings	Keywords
Yuan et al. [9]	LR-HR paired images are not available so accurate kernel estimation and supervised learning are not achievable	Without paired data, we can use unsupervised learning	Cycle-in-Cycle [31, 32] network	NTIRE2018 Super Resolution Challenge	Comparable results are obtained using the proposed unsupervised technique and cutting-edge supervised models	Impaired data, data augmentation, Image synthesizing
Lin et al. [10]	Most of the model use supervised GAN for generating SR	To generating the SR image by using unsupervised way	Deep unsupervised learning SR based on GANs	Berkeley segmentation dataset BSDS500, SET5, SET14	In a subjective visual evaluation, the approach produces sharp edges with uncommon artefacts that are closer to the real world	Deep unsupervised learning, Sub-pixel convolution, Regularization

Table 3. Performance metrics and studies in which they are used

Performance	Metrics Studied by Authors
PSNR	Mahapatra et al. [1] (2017) Tzu-An Song et al. [3] (2020) Gu et al. [4] (2020) Ledig et al. [5] Haoyu Ren and Kheradmand et al. [6] Dongyang Zhang et al. [7] Dwarikanath Mahapatra et al. [8] Yuan Yuan et al. [9] Guimin Lin et al. [10] Yuhua Chen et al. [11] Xintao Wang et al. [12] Jiabo Ma et al. [15] Wenlong Zhang et al. [16] Luis Salgueiro Romero et al. [17] Irina Sánchez et al. [18] Nathanael Carraz Rakotonirina et al. [19] Xinyang Bing et al. [20] Quentin Delannoy et al. [21] Qian Huang et al. [22] Kalpesh Prajapati et al. [23] Lijun Zhao et al. [24] Anwesh Kabiraj et al. [25]

(*continued*)

Table 3. (*continued*)

Performance	Metrics Studied by Authors
SSIM	Mahapatra et al. [1] (2017)
	Tzu-An Song et al. [3] (2020)
	Guimin Lin et al. [4] (2020)
	Ledig et al. [5]
	Haoyu Ren and Kheradmand et al. [6]
	Dongyang Zhang et al. [7]
	Dwarikanath Mahapatra et al. [8]
	Yuan Yuan et al. [9]
	Guimin Lin et al. [10]
	Xu et al. [13]
	Jiabo Ma et al. [15]
	Romero et al. [17]
	Irina Sánchez et al. [18]
	Nathanael Carraz Rakotonirina et al. [19]
	XINYANG BING et al. [20]
	Delannoy et al. [21]
	Qian Huang et al. [22]
	Kalpesh Prajapati et al. [23]
	Lijun Zhao et al. [24]
	Anwesh Kabiraj et al. [25]
MSE	Mahapatra et al. [1] (2017)
	Tzu-An Song et al. [3] (2020)
	Guimin Lin et al. [4] (2020)
	Ledig et al. [5]
	Dwarikanath Mahapatra et al. [8]
	Yuhua Chen et al. [11]
	Xintao Wang et al. [12]
	Xu et al. [13]
	Corley et al. [14]
	Jiabo Ma et al. [15]
	XINYANG BING et al. [20]

4 Discussion and Conclusion

Here we utilize all the previous research which has been under this topic and make a compiled report. It is covering almost all domains in which SRGAN is applied or will going to apply. In Table 1 when we are rigorously analyzing the research paper, we identify the problem, for that problem, identify the solution, benchmark dataset, and contribution of the author in that area. By just looking at the keywords, anyone can figure that what authors did in their research. After figuring out this task we have evaluated that most of the researchers used mainly two metrics for analyzing the quality of the image as shown in Table 2 first is the SSIM and another is the PSNR value. But the very least playing evaluation metric is the MSE because, sometimes during the processing of the image, the values of MSE of the base image and the target images get overshot.

Abbreviations

SR	Super-Resolution
HR	High-Resolution

LR	Low-Resolution
CNN	Convolutional Neural Networks
SNNC	Sub-pixel Nearest Neighbor Convolution
P-GAN	Progressive Generative Adversarial Networks
PSNR	Peak-Signal-to-Noise Ratios
SISR	Single Image Super-Resolution
MSE	Mean Squared Error
DRN	Deep Residual Network
SRGAN	Super-Resolution GAN
SISR	Singe Image Super-Resolution
PET	Positron Emission Tomography
SSSR	Self-Supervised SR
CT	Computer Tomography
ERM	Empirical Risk Minimization
MedSRGAN	Medical Images SR using GAN

References

1. Mahapatra, D., Bozorgtabar, B., Hewavitharanage, S., Garnavi, R.: Image super resolution using generative adversarial networks and local saliency maps for retinal image analysis. In: Descoteaux, M., Maier-Hein, L., Franz, A., Jannin, P., Collins, D.L., Duchesne, S. (eds.) MICCAI 2017. LNCS, vol. 10435, pp. 382–390. Springer, Cham (2017). https://doi.org/10.1007/978-3-319-66179-7_44

2. Demiray, B.Z., Sit, M., Demir, I.: D-SRGAN: DEM super-resolution with generative adversarial networks. SN Comput. Sci. **2** (2021). https://doi.org/10.1007/s42979-020-00442-2

3. Song, T.A., Chowdhury, S.R., Yang, F., Dutta, J.: PET image super-resolution using generative adversarial networks. Neural Netw. **125**, 83–91 (2020). https://doi.org/10.1016/j.neunet.2020.01.029

4. Gu, Y., et al.: MedSRGAN: medical images super-resolution using generative adversarial networks. Multimed. Tools Appl. **79**, 21815–21840 (2020). https://doi.org/10.1007/s11042-020-08980-w

5. Ledig, C., et al.: Photo-realistic single image super-resolution using a generative adversarial network. In: CVPR, vol. 2, p. 4 (2017). https://ieeexplore.ieee.org/abstract/document/8099502

6. Ren, H., Kheradmand, A., El-Khamy, M., Wang, S., Bai, D., Lee, J.: Real-world super-resolution using generative adversarial networks. In: IEEE Conference on Computer Vision and Pattern Recognition Work, 2020-June, pp. 1760–1768 (2020). https://doi.org/10.1109/CVPRW50498.2020.00226

7. Zhang, D., Shao, J., Hu, G., Gao, L.: Sharp and real image super-resolution using generative adversarial network. In: Liu, D., Xie, S., Li, Y., Zhao, D., El-Alfy, ES. (eds.) ICONIP 2017. LNCS, vol. 10636. Springer, Cham (2017). https://doi.org/10.1007/978-3-319-70090-8_23

8. Mahapatra, D., Bozorgtabar, B., Garnavi, R.: Image super-resolution using progressive generative adversarial networks for medical image analysis. Comput. Med. Imaging Graph. **71**, 30–39 (2019). https://doi.org/10.1016/j.compmedimag.2018.10.005

9. Yuan, Y., Liu, S., Zhang, J., Zhang, Y., Dong, C., Lin, L.: Unsupervised image super-resolution using cycle-in-cycle generative adversarial networks. In: IEEE Conference on Computer Vision and Pattern Recognition Work, 2018-June, pp. 814–823 (2018). https://doi.org/10.1109/CVPRW.2018.00113

10. Lin, G., et al.: Deep unsupervised learning for image super-resolution with generative adversarial network. Sig. Process. Image Commun. **68**, 88–100 (2018). https://doi.org/10.1016/j.image.2018.07.003

11. Kushwaha, R.S., Rakhra, M., Singh, D., Singh, A.: An overview: super-image resolution using generative adversarial network for image enhancement, pp. 1243–1246. IEEE, Uttar Pradesh, India (2022). https://doi.org/10.1109/IC3I56241.2022.10072862

12. Wang, X., Yu, K., Wu, S., Gu, J., Liu, Y.: ESRGAN : Enhanced Super-Resolution Generative Adversarial Networks, pp. 1–16

13. Xu, L., Zeng, X., Huang, Z., Li, W., Zhang, H.: Low-dose chest X-ray image super-resolution using generative adversarial nets with spectral normalization. Biomed. Sig. Process. Control **55**, 101600 (2020). https://doi.org/10.1016/j.bspc.2019.101600

14. Corley, I.A., Huang, Y.: Dataset, V, pp. 4–7 (2018)

15. Ma, J., et al.: PathSRGAN : multi-supervised super- resolution for cytopathological images using generative adversarial network. IEEE Trans. Med. Imaging **39**, 2920–2930 (2020)

16. Qiao, Y., Zhang, W., Liu, Y.: RankSRGAN : generative adversarial networks with ranker for image super-resolution university of Chinese academy of sciences. In: Proceedings of the IEEE/CVF International Conference on Computer Vision, pp. 3096–3105 (2019)

17. Romero, L.S., Marcello, J., Vilaplana, V.: Super-resolution of Sentinel-2 imagery using generative adversarial networks. Remote Sens. **12**, 1–25 (2020). https://doi.org/10.3390/RS12152424

18. Sanchez, I., Vilaplana, V.: Brain MRI super-resolution using 3D generative adversarial networks, pp. 1–8 (2018)

19. Rakotonirina, N.C., Rasoanaivo, A.: ESRGAN+ : Further Improving Enhanced Super-Resolution Generative Adversarial Network, pp. 3637–3641 (2020). Laboratoire d'Informatique et Mathématiques, Université d'Antananarivo, Madagascar

20. Bing, X., Zhang, W., Zheng, L., Zhang, Y.: Medical image super resolution using improved generative adversarial networks. IEEE Access **7**, 145030–145038 (2019). https://doi.org/10.1109/ACCESS.2019.2944862

21. Delannoy, Q., et al.: SegSRGAN: super-resolution and segmentation using generative adversarial networks—Application to neonatal brain MRI. Comput. Biol. Med. **120**, 103755 (2020). https://doi.org/10.1016/j.compbiomed.2020.103755

22. Huang, Q., Li, W., Hu, T., Tao, R.: Hyperspectral image super-resolution using generative adversarial network and residual learning. In: Proceedings of ICASSP, International Conference on Acoustics, Speech, and Signal Processing, 2019-May, pp. 3012–3016 (2019). https://doi.org/10.1109/ICASSP.2019.8683893

23. Prajapati, K., et al.: Unsupervised single image super-resolution using cycle generative adversarial network. Commun. Comput. Inf. Sci. **1382**, 359–370 (2021). https://doi.org/10.1007/978-3-030-71711-7_30

24. Zhao, L., Bai, H., Liang, J., Zeng, B., Wang, A., Zhao, Y.: Simultaneous color-depth super-resolution with conditional generative adversarial networks. Pattern Recogn. **88**, 356–369 (2019). https://doi.org/10.1016/j.patcog.2018.11.028

25. Kabiraj, A., Pal, D., Ganguly, D., Chatterjee, K., Roy, S.: Number plate recognition from enhanced super-resolution using generative adversarial network. Multimed. Tools Appl. (2022). https://doi.org/10.1007/s11042-022-14018-0

26. Super-resolution of remote sensing images based on transferred generative adversarial network, pp. 1148–1151 (2018)

27. Lv, B., Liu, Y., Zhang, S., Zeng, H., Zhu, G.: Super Resolution with Generative Adversarial Networks (2018)

28. Kasem, H.M., Hung, K., Jiang, J.: Spatial transformer generative adversarial network for robust image super-resolution. IEEE Access. **7**, 182993–183009 (2019). https://doi.org/10.1109/ACCESS.2019.2959940

29. Das, V., Dandapat, S., Bora, P.K.: Unsupervised super-resolution of OCT images using generative adversarial network for improved age-related macular degeneration diagnosis **20**, 8746–8756 (2020)

30. Adate, A., Tripathy, B.K.: Super-Resolution Techniques with Generative Adversarial Networks. Springer, Singapore. https://doi.org/10.1007/978-981-13-1592-3

31. Kim, G., et al.: Unsupervised Real-World Super Resolution with Cycle Generative Adversarial Network and Domain Discriminator

32. Zhang, Y., Liu, S., Dong, C., Zhang, X.: Multiple cycle-in-cycle generative adversarial networks for unsupervised image super-resolution **29**, 1101–1112 (2020).

33. Wang, H., Wu, W., Su, Y., Duan, Y., Wang, P.: Image Super-Resolution Using a Improved Generative Adversarial Network, pp. 23–26 (2015)

34. Ma, Y., Liu, K., Xiong, H., Fang, P., Li, X., Chen, Y.: Medical image super-resolution using a relativistic average generative adversarial network. Nucl. Instrum. Methods Phys. Res. A. **992**, 165053 (2021). https://doi.org/10.1016/j.nima.2021.165053

35. Chen, B., Liu, T., Liu, K., Liu, H., Pei, S.: Image Super-Resolution Using Complex Dense Block on Generative Adversarial Networks, Department of Electrical Engineering, National Chung Hsing University Graduate Institute of Communication Engineering, National Taiwan University, pp. 2866–2870 (2019)

36. Mahapatra, D., Bozorgtabar, B., Hewavitharanage, S.: Saliency Maps and Generative Adversarial

37. Daihong, J.: Multi-scale generative adversarial network for image super-resolution. Soft. Comput. **26**, 3631–3641 (2022). https://doi.org/10.1007/s00500-022-06822-5

38. Adversarial Networks: Small Object Detection in Remote Sensing Images Based on Super-Resolution with Auxiliary Generative Adversarial Networks (2020). https://doi.org/10.3390/rs12193152

39. Cai, J., Meng, Z., Ho, C.M.: Residual Channel Attention Generative Adversarial Network for Image Super-Resolution and Noise Reduction

40. He, J., Zheng, J., Shen, Y., Guo, Y., Zhou, H.: Neurocomputing facial image synthesis and super-resolution with stacked generative adversarial network **402**, 359–365 (2020). https://doi.org/10.1016/j.neucom.2020.03.107

Stance Detection in Manipuri Editorial Article Using CRF

Pebam Binodini[1]([✉]), Kishorjit Nongmeikapam[2], and Sunita Sarkar[1]

[1] Computer Science and Engineering, Assam University, Silchar, India
tambipebam@gmail.com
[2] Computer Science and Engineering, IIIT Imphal, Imphal, India
kishorjit@iiitmanipur.ac.in

Abstract. The primary purpose of stance detection is to categorize the author's attitude as favourable to, antagonistic towards, or neutral to a specific target. Most stance detection work available is on short text emphasizing the English Language and less on targeting long text, especially in low-resource languages like Manipuri. Based on the article "Citizenship Amendment Bill (CAB)" this experiment performs stance detection in Manipuri. Here, a model is created through using Conditional Random Field (CRF) to decide each word's polarity, and the article's stance is then decided. The best features recorded for identifying a word's polarity is given in the form of Precision: 61.78%, Recall: 63.23%, and F_Score: 62.49%. The model gives the result with a mean accuracy of 56.66%.

Keywords: Stance · CRF · Polarities · Meitei Mayek · Low-Resource Language · Editorial Article

1 Introduction

People tend to express their feelings on newspaper platforms or social media. They share their views and feelings on different topics in various fields like social issues, politics, movies, sports, etc. People tend to take a stance on a particular topic during this process. Public opinion analysis on different issues relies on stance detection. Stance detection is a critical element of analysis studies that measure public sentiment on media platforms, especially regarding political and social issues. People often express opposing opinions on specific points, making these topics controversial. Detecting a text's Stance is the same as predicting its Stance (such as whether it is in support of or against a particular target). Stance conveys the author's perspective and assessment of a particular proposition. It has shown to be a useful tool for determining whether rumours are true (rumour classification) and for spotting fake news [1]. More knowledge can be gained about user perceptions of the target entities by mining comments, tweets, or posts on social media platforms. This has many practical applications in marketing research and election campaigning. Gathering consumer feedback on a product is known as market research, which is done in order to take strategic business actions related to the product. The public's response on social media aids the government in assessing its

P. Das et al. (Eds.): AMRIT 2023, CCIS 1954, pp. 187–197, 2024.
https://doi.org/10.1007/978-3-031-47221-3_17

new election-related policies. Information retrieval, opinion mining, text entailment, and other tasks use the stance detection system. More annotated data must be available to handle stance detection in longer texts and languages other than English because many of the benchmark stance detection dataset focuses on short text snippets, particularly tweets [1, 2]. As far as the authors knowledge, no documentation on stance detection is recorded in Manipuri which is eighth scheduled language listed in the Indian Constitution and also is an official language of the Indian Government. Manipuri being a low-resource language, is very challenging to work with due to its highly agglutinative nature [3]. The people of Manipur use two scripts, one Bengali, which is a borrowed one, and the other script of Manipuri which is Meitei Mayek. The experiment is carried out on articles written in Meitei Mayek. To perform stance detection first it should have a topic for discussion which is "Citizenship Amendment Bill" for the experiment. And second, it should have a text or comment from an author which are Editorial articles collected from a daily local newspaper[1]. Editorial article is a persuasive text that provides a rich source of discourse analysis on a specific event or thing. It is critical to study an actual event or a subject that has emerged unexpectedly before society and has become somewhat controversial. It can influence decision-makers, sway public opinion on a large scale, and persuade people. As a result, the Stance Detection of editorials plays a critical role in public or organizational decision-making.

The paper is arranged as: Related Work in Sect. 2, Conditional Random Field in Sect. 3, System Methodology in Sect. 4, Experiment and Evaluation in Sect. 5, Validations in Sect. 6, Conclusions in Sect. 7.

2   Related Work

Stance recognition task in NLP is well known and extensively studied. The objective is to deduce whether the author of the text supports, opposes, or is neutral in regards to the target given [2]. Recently, it has been gaining attention as a subject. Although distinctive, it is closely related to the sentiment analysis task. The author's [4] did some of the early work. They explore the use of sentiment and arguing features. For the case study, they experiment with ideological debates. From a corpus manually annotated, they construct an arguing lexicon. Later, [5] notes that the classification outcomes for a given topic using lexical and contextual features are significantly superior to the finest features without any contextual information when analyzing the dialogic arrangement of debates. Many jobs, such as detecting fake news [6], claim validation [7], and argument search [8], now rely on stance detection. The dataset [2] is an English-annotated dataset with 5 collections of targets. There are also tweets dataset available for other language such as Spanish and Catalan [9], French [10]. The multilingual dataset of German version known as X-stance includes a selection of political questions as well as the candidates' responses to those questions. This dataset is also accessible in French and Italian [11]. For the English news articles, two different datasets are available in the field of journalism. The Emergent dataset, which is one component of fact-checking, includes claims as well as news stories that are important to those charges [12]. In a similar vein, the

---

[1] Https://sanaleibak.in/archives/.

dataset of news stories and the various places on a headline that are included in the Fake News Challenge detection, sometimes commonly referred to as FNC-1, may be found here[2]. Emotion recognition datasets focus mostly on brief text fragments, such as tweets, headlines of news [13], conversations of text [14] or words in fairy tales are the focus of the majority of emotion detection datasets [15]. Emotional flow has been explored in books and film synopsis [16], but not in journalistic content. For the purpose of improving stance recognition, several research methodologies place a greater focus on the utilisation of sentiment data, and more specifically, the text polarity. In point of fact, the SemEval annual workshop arranged a group project called "stance identification" and offered a dataset of tweets that had been tagged with sentiment data. Results on whether sentiment data enhances stance detection are conflicting. Some studies claim that using such information is beneficial [17, 18], while others claim that the tasks are unrelated. For instance, a text with an attitude could disagree with a particular assertion [1, 19].

3   Conditional Random Field (CRF)

CRF is mostly used in sequential task prediction problems where neighboring state or the contextual information plays an essential part in predicting the current state. Some applications of CRF are tagging Parts of Speech, Recognizing Named Entity, Object detection problems, etc. CRF belongs to the category of discriminative models. CRF evaluates the conditional probabilities of values from undirected graphs on additional selected input nodes. It is a supervised machine learning and testing technique in which the system will be given a training set in order to enable learning. When the observed sequence is $Y = (y_1, y_2, y_3, \ldots, y_T)$ the conditional probability of a state sequence, $X = (x1, x2, x3, \ldots, xT)$, is given in the following:

$$P\left(\frac{Y}{X}\right) = \frac{1}{z_x} \exp\left(\sum_{t=1}^{T} \sum_{k} \lambda_k f_k (y_{t-1}, y_t, X, t)\right) \tag{1}$$

In Eq. 1, $f_k(y_{t-1}, y_t, X, t)$ is known as feature function. k, learnt weight and associates with f_k that learns through training. T is the time taken to calculate the function. Feature functions ranges from minus infinity to plus infinity and more often than not, they are binary. Z_x is called as normalization factor and this needs to be calculated to make the sum of all state sequences probability to 1. Z_x is given by:

$$Z_x = \sum_y \exp \sum_{t=1}^{T} \sum_k \lambda_k f_k (y_{t-1}, y_t, X, t) \tag{2}$$

This is evaluated in HMM (Hidden Markov Model) and can be derived by dynamic programming very easily. Because the conditional likelihood P(Y/X) is defined by CRF. The conditional likelihood of the data or state sequence should be maximized as the goal of parameter learning.

$$\sum_{i=1}^{N} \log P\left(y^i \big/ x^i\right) \tag{3}$$

$\{(x^i, y^i)\}$ denotes the data labelled. In order to regularize the process of training on the data, Gaussian is used before λ. All λ's will have a specific set of optional values if λ ~ N $(0, \rho^2)$, the function of the objective takes the shape of concave and is given by

$$\sum_{i-1}^{N} logP\left(y^i/x^i\right) - \sum_{k} \frac{\lambda_i^2}{2\rho^2} \tag{4}$$

4 System Methodology

4.1 Data Collection and Preprocessing

The articles are collected from a prominent local daily newspaper for the experiment and the articles collected are mostly on the topic "Citizenship Amendment Bill." The experiment is performed on thirty articles. A linguistic expert is engaged in examining any spelling mistakes, syntax, or necessary changes needed. The articles are then tokenized and each token are segmented to identify the affixes. Each token is represented as a sentence. The features whichever felt necessary are selected and white space separates every word in the row. The feature combination and selection are made possible by the template file which is present in the CRF++. Section B discusses the list of features used in the experiment. Figure 1 gives the block diagram of the model.

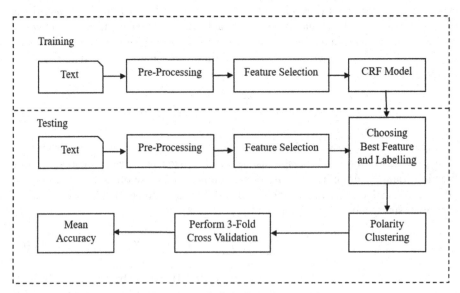

Fig. 1. Block Diagram of the model

4.2 Feature List

Following are the list of features used in the experiment:

Adjoining Words: The word(s) that come before or after the current word are crucial in determining speech tagging because they have a significant impact on determining the parts of speech of the current term.

Adjoining Stem Words: Parts of speech mainly focus on a particular word's preceding and following stemmed words. The stemming algorithm described in [20] is used. This marks as an essential feature.

Acceptable Standard Suffixes: According to the information presented in [20], suffixes plays a significant part in determining the PoS in a very agglutinative language like Manipuri. A word in Manipuri can have at the maximum ten suffixes as per the report, these standard suffixes are identified, and serves as a feature in the experiment.

Acceptable Standard Prefixes: The prefixes described in [20] is used and is essential in forming the parts of speech of a word.

Word Length: Word length is calculated which is number of letters in a word. This feature is essential part. If the word length is less than 4, this column is set to 1 otherwise it is set to 0.

Frequency of the Word: Frequency also plays a vital role as determiners, pronouns and conjunctions occur frequently. If the frequency of the word is less than fifty it is set to 0 else it is set to 1.

Digits: Digits represent quantities, dates, and monetary values. Therefore, the digit feature is essential. If the token is a digit, it marks 1otherwise 0.

Special Symbols: Signs like $,%, etc. have meaning in the text, so it is set to 1 if they are present and 0 if not.

PoS: This feature has been manually assigned by linguistic experts. The verbs, adverbs, adjectives, and nouns in the sentence can represent the polarity of the words in it in a significant way.

Word Polarity: This feature also has been manually tagged. This plays a very important part in the experiment. The word is indicated as "pos" if it is positive, "neg" if it is negative, and "neu" if it is neutral. If the word is either conjunction or pronouns it is assigned or tagged with 0.

5 Experiment and Evaluation

For the experiment, CRF++, 0.58 package is used. This package is based on C++ and its main purpose is for research work. It is free to download and is used for division or labelling sequential data. For the experiment two standard files are created one known as training and the other known as testing. For CRF++ to work correctly, the training as well as the testing files must follow a specific format. Training and testing file should typically contain multiple tokens. A token also includes numerous columns with fixed numbers. Template files used these columns, offering thorough information about the option configuration. Each line displays the token, and white space separates the columns. The template files enable the feature combinations used in the experiment. A sample of the input training file is given in Fig. 2

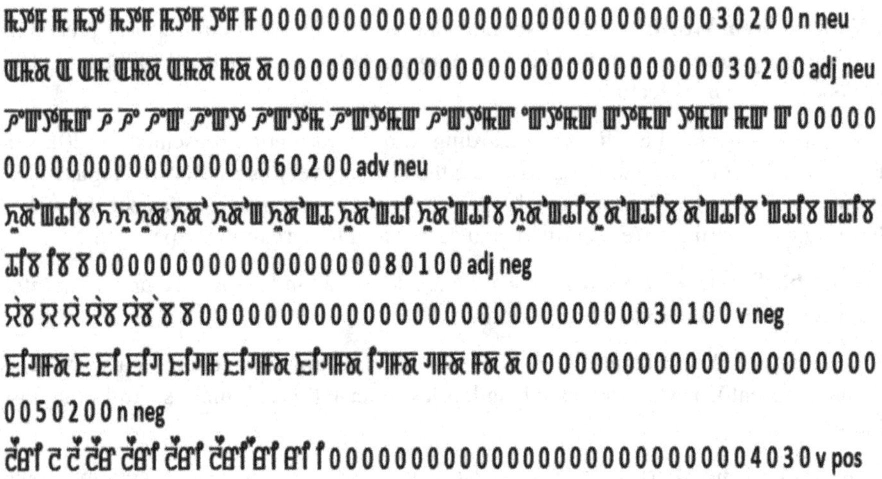

ਜ਼ੋਫ ਫ ਜ਼ੋ ਜ਼ੋਫ ਜ਼ੋਫ ਯੋਫ ਫ 0 0 0 0 0 0 0 0 0 0 0 0 0 0 0 0 0 0 0 0 0 0 0 0 0 0 0 3 0 2 0 0 n neu

ਘੜਾ ਘ ਘਾ ਘੜਾ ਘੜਾ ਜ਼ੜਾ ੜਾ 0 0 0 0 0 0 0 0 0 0 0 0 0 0 0 0 0 0 0 0 0 0 0 0 0 0 0 3 0 2 0 0 adj neu

ਯੰੁਯੋਜ਼ੋ ਯ ਯ ਯੰ ਯੰੁਯੋ ਯੰੁਯੋਜ਼ ਯੰੁਯੋਜ਼ੋ ਯੰੁਯੋਜ਼ੋ ੰੁਯੋਜ਼ੋ ੁਯੋਜ਼ੋ ਯੋਜ਼ੋ ਜ਼ੋ ੋ 0 0 0 0 0
0 0 0 0 0 0 0 0 0 0 0 0 0 0 6 0 2 0 0 adv neu

ਨ੍ਹੇੱਘੜਾਂ ਨ ਨੇ ਨ੍ਹੇ ਨ੍ਹੇੱ ਨ੍ਹੇੱਘ ਨ੍ਹੇੱਘੜ ਨ੍ਹੇੱਘੜਾ ਨ੍ਹੇੱਘੜਾਂ ੇੱਘੜਾਂ ੱਘੜਾਂ ਘੜਾਂ
ੜਾਂ ਾਂ ਂ 0 0 0 0 0 0 0 0 0 0 0 0 0 0 0 0 0 0 8 0 1 0 0 adj neg

ਖੜੇ ਖ ਖੇ ਖੜੇ ਖੜੇੱ ੜੇੱ 0 0 0 0 0 0 0 0 0 0 0 0 0 0 0 0 0 0 0 0 0 0 0 0 0 0 3 0 1 0 0 v neg

ਈਗਾਜ਼ ਈ ਈਾ ਈਾਗ ਈਾਗਾ ਈਾਗਾਜ਼ ਈਾਗਾਜ਼ ਾਗਾਜ਼ ਗਾਜ਼ ਜ਼ ੋ 0 0 0 0 0 0 0 0 0 0 0 0 0 0 0 0 0 0 0 0
0 0 5 0 2 0 0 n neg

ਚੰੈਾ ਚ ਚੰ ਚੰੈ ਚੰੈਾ ਚੰੈਾ ਚੰੈਾਾ ੰੈਾ ੈਾ ਾ 0 0 0 0 0 0 0 0 0 0 0 0 0 0 0 0 0 0 0 0 0 0 0 0 4 0 3 0 v pos

Fig. 2. Sample Input

5.1 Model File

Model file is produced after training. The CRF tool creates this model file in preparation for use in the testing procedure. The CRF-learned file created after training is the model file. The model file contains comprehensive information from both the template file and training file, so there is no need to reuse them.

5.2 Choosing Best Feature and Labelling

The features are manually chosen based on what is essential and carry experiments to determine the best features. To determine the word's polarity, different feature combinations are made. The combinations that can identify the word's polarity correctly will be chosen as the best feature combinations. Various feature notations used in the experiment is given in Table 1. Each feature combination efficiency is calculated using Recall, Precision and F_Score value. The combination which gives the highest Precision, Recall and F-Score value will be the best feature combination. Various feature combinations and their Precision, Recall and F-Score values resulted from the experiment is given in Table 2. The formula for finding the Precision, Recall and F-Score value are given in Eq. 5, 6 and 7 below:

$$\text{Precision} = \frac{\text{total number of correct answers provided by the system}}{\text{the number of correct answers found in the text}} \tag{5}$$

$$\text{Recall} = \frac{\text{total number of correct answers provided by the system}}{\text{the number of answers provided by the system}} \quad (6)$$

$$\text{F\_Score} = \frac{(\beta^2 + 1)PR}{\beta^2 P + R} \quad (7)$$

The value of β is taken as 1 and Precision and Recall have equal weights. The experiment has a few issues because the Manipuri language is so distinctive. In Manipuri, word categorization could be more precise. The verbs are also categorized as bound. Although there is usually no doubt between the noun class and the verb class, the line between the noun class and the adjective class is frequently hazy. Because a word may be a noun structurally but an adverb contextually, it can be difficult to tell them apart. Incorrect tagging may result from a prefix that is also a portion of the root. Noun morphology is simpler than verbal morphology. A complete word can occasionally be created by combining two different words.

Table 1. Notations of Feature

Notation	Meaning
SW	Surrounding words
StW[−K, +L]	Stem words from k^{th} left to l^{th} right positions
Pref	Word prefixes
Suff	Word suffixes
Dg	Digit feature
Freq	Word frequency
Len	Word length
Sym	Special symbol
POS[−K, +L]	Parts of speech from k^{th} left to l^{th} right positions
P	Polarity of the word

Table 2. Experimental Result

Feature combination	Recall (%)	Precision (%)	F-Score (%)
SW, StW[−2, +2], Pref, Suff, Dg, Freq, Len, Sym, POS[−2, +2], P	63.23	61.78	62.49
SW, StW[−3, + 2], Pref, Suff, Dg, Freq, Len, Sym, POS[−3, +2], P	58.32	57.01	57.66
SW, StW[−3, + 3], Pref, Suff, Dg, Freq, Len, Sym, POS[−3, +3], P	56.76	55.67	56.209
SW, StW[−4, + 3], Pref, Suff, Dg, Freq, Len, Sym, POS[−4, +3], P	42.04	41.97	41.92
SW, StW[−4, + 4], Pref, Suff, Dg, Freq, Len, Sym, POS[−4, +4], P	33.21	29.34	31.15

5.3 Best Result

The best feature combination observed for identifying the polarity of each word is given in Table 3.

Table 3. Best Feature Combination

Feature	Recall (%)	Precision (%)	F-Score (%)
SW, StW[−2, +2], Pref, Suff, Dg, Freq, Len, Sym, POS[−2, + 2], P	63.23	61.78	62.21

5.4 Test File

This best feature resulted above is used for labelling the polarity in test file in testing process. The test file has fixed columns and same field as the training file. A file with an additional polarity column that the system automatically labels is created with the testing output.

5.5 Polarity Clustering

Here, the polarity of each word is determined, and the sum total for the positive, negative, and neutral groups are calculated. The document's stance is determined by the three types of polarities' highest sum. If the number of positive polarities is highest among the three, then the article stands for "for" the article towards the target. If the number of negative polarities is highest, it stands for "against" the target, and if the total number of neutral is highest, it stands for "neutral" towards the target. To find the system mean-accuracy, 3-fold cross validation is performed.

6 Validation

Depending on the message the articles convey, the expert annotates each article used in the experiment as positive, negative, or neutral. The experiment uses K-fold cross-validation to calculate the system's mean accuracy. K-fold cross validation is a type of cross-validation where machine learning models are tested on a small sample of data. How many divisions from a specific data sample should be made in the procedure is determined by one parameter, k. So, the name k-fold cross-validation. It is primarily used in applied machine learning to assess how well a model performs when applied to new data. To make predictions on data that was not used during the model's training, this entails using a small sample to estimate how the model is expected to perform. It is a well-liked technique because it is straightforward to comprehend and frequently yields a less biased or optimistic estimate of the model skill than other techniques. For the experiment, the training data is divided into 3 sets, set M, set N and set O. For set M: Training is done on combinations of set N and set O and the test set sum is determined on set M. For set N: Training is done on combinations of set M and set O and the test set sum is determined on set N. For set O: Training is done on combinations of set M and set N and the test set sum is determined on set O. All the articles used in the experiment are manually annotated by experts. The following Table 4 is the result obtained for the Stance Detection on "Citizenship Amendment Bill".

Table 4. .

Article combination among set M, set N and set O	Test article set	Number of articles correctly detected
M and N	O	6
N and O	M	6
O and M	N	5

Here, in the above table, set M, set N and set O consists of 10 articles each and each article consists of approximately 550 words. The experiment records a mean-accuracy of 56.66%.

7 Conclusion

This work presents stance detection for Manipuri Editorial article at the document level. The job necessitates manual feature selection. Various feature combinations has been performed. The best feature combination to identify the polarity of a word in an article records Precision of 61.78%, Recall of 63.23%, and F-score of 62.49%. And gives system mean-accuracy of 56.66%. There is a great deal of room for development by learning more on how to enhance stance detection methods. In addition, the system can be more effective at determining an article's stance by using cutting-edge techniques.

References

1. Mohammad, S., Kiritchenko, S., Sobhani, P., Zhu, X., Cherry, C.: Semeval-2016 task 6: detecting stance in tweets. In: Proceedings of the 10th International Workshop on Semantic Evaluation (SemEval-2016), pp. 31–41 (2016b)
2. Hanselowski, A., et al.: A retrospective analysis of the fake news challenge stance-detection task. In: Proceedings of the 27th International Conference on Computational Linguistics, pp. 1859–1874. Association for Computational Linguistics, Santa Fe (2018)
3. Nongmeikapam, K., Shangkhunem, T., Chanu, N.M., Singh, L.N., Salam, B., Bandyopadhyay, S.: CRF based name entity recognition (NER) in Manipuri: a highly agglutinative Indian language. In: Proceedings - 2011 2nd National Conference on Emerging Trends and Applications in Computer Science, NCETACS-2011, pp. 92–97 (2011)

4. Somasundaran, S., Wiebe, J.: Recognizing stances in ideological on-line debates. In: Proceedings of the NAACL HLT 2010 Workshop on Computational Approaches to Analysis and Generation of Emotion in Text, pp. 116–124 (2010)

5. Anand, P., Walker, M., Abbott, R., Tree, J.E.F., Bowmani, R., Minor, M.: Cats rule and dogs drool!: Classifying stance in online debate. In: Proceedings of the 2nd Workshop on Computational Approaches to Subjectivity and Sentiment Analysis (WASSA 2011), pp. 1–9 (2011)

6. Sharma, S., Saraswat, M., Dubey, A.K.: Fake news detection using deep learning. In: Villazón-Terrazas, B., Ortiz-Rodríguez, F., Tiwari, S., Goyal, A., Jabbar, M.A. (eds.) KGSWC 2021. CCIS, vol. 1459, pp. 249–259. Springer, Cham (2021). https://doi.org/10.1007/978-3-030-91305-2_19

7. Popat, K., Mukherjee, S., Strötgen, J., Weikum, G.: Where the truth lies: explaining the credibility of emerging claims on the web and social media. In: Proceedings of the 26th International Conference on World Wide Web Companion, pp. 1003–1012 (2017)

8. Stab, C., Miller, T., Gurevych, I.: Cross-topic argument mining from heterogeneous sources using attention-based neural networks. arXiv preprint arXiv:1802.05758 (2018)

9. Zotova, E., Agerri, R., Rigau, G.: Semi-automatic generation of multilingual datasets for stance detection in Twitter. Expert Syst. Appl. **170** (2021)

10. Cignarella, A.T., Lai, M., Bosco, C., Patti, V., Paolo, R.: Sardistance @ EVALITA2020: overview of the task on stance detection in Italian tweets. In: Proceedings of the 2020 Evaluation Campaign of Natural Language Processing and Speech Tools for Italian (EVALITA). CEUR-WS (2020, online)

11. Vamvas, J., Sennrich, R.: X-stance: a multilingual multi-target dataset for stance detection. In: Proceedings of the 5th Swiss Text Analytics Conference (SwissText) & 16th Conference on Natural Language Processing (KONVENS), Zurich, Switzerland (2020)

12. Ferreira, W., Vlachos, A.: Emergent: a novel data-set for stance classification. In: Proceedings of the 2016 Conference of the North American Chapter of the Association for Computational Linguistics: Human Language Technologies. The 15th Annual Conference of the North American Chapter of the Association for Computational Linguistics: Human Language Technologies, 12–17 June 2016. ACL, San Diego (2016). ISBN 978-1-941643-91-4

13. Mohammad, S., Kiritchenko, S., Sobhani, P., Zhu, X., Cherry, C.: A dataset for detecting stance in tweets. In: Proceedings of the Tenth International Conference on Language Resources and Evaluation (LREC 2016), pp. 3945–3952. European Language Resources Association (ELRA), Portorož (2016)

14. Strapparava, C., Mihalcea, R.: SemEval-2007 task 14: affective text. In: Proceedings of the Fourth International Workshop on Semantic Evaluations (SemEval-2007), pp. 70–74, Association for Computational Linguistics, Prague (2007)

15. Volkova, E.P., Mohler, B.J., Meurers, D., Gerdemann, D., Bülthoff, H.H.: Emotional perception of fairy tales: achieving agreement in emotion annotation of text. In: CAAGET 2010: Proceedings of the NAACL HLT 2010 Workshop on Computational Approaches to Analysis and Generation of Emotion in Text, June 2010

16. Kar, S., Maharjan, S., López-Monroy, A.P., Solorio, T.: MPST: a corpus of movie plot synopses with tags. In: Proceedings of the Eleventh International Conference on Language Resources and Evaluation (LREC 2018). European Language Resources Association (ELRA), Miyazaki (2018)

17. Li, Y., Caragea, C.: Multi-task stance detection with sentiment and stance lexicons. In: Proceedings of the 2019 Conference on Empirical 75 Methods in Natural Language Processing and the 9th International Joint Conference on Natural Language Processing (EMNLP-IJCNLP), pp. 6299–6305. Association for Computational Linguistics, Hong Kong (2019)

18. Sun, Q., Wang, Z., Li, S., Zhu, Q., Zhou, G.: Stance detection via sentiment information and neural network model. Front. Comput. Sci. **13**(1), 127 (2019)
19. Aldayel, A., Magdy, W.: Assessing sentiment of the expressed stance on social media. In: Weber, I., Darwish, K.M., Wagner, C., Zagheni, E., Nelson, L., Aref, S., Flöck, F. (eds.) SocInfo 2019. LNCS, vol. 11864, pp. 277–286. Springer, Cham (2019). https://doi.org/10.1007/978-3-030-34971-4_19
20. Nongmeikapam, K., Bandyopadhyay, S., Salam, B., Romina, M., Chanu, N.M.: A light weight Manipuri stemmer. In: The Proceedings of National Conference on Indian Language Computing (NCILC), Cochin, India (2011)

Deep Learning Based Software Vulnerability Detection in Code Snippets and Tag Questions Using Convolutional Neural Networks

Anurag Khanra$^{(\boxtimes)}$ (ID), Arvind Krishna (ID), L. H. Jeevan Samrudh (ID),
Rahul D. Makhija (ID), and V. R. Badri Prasad

Department of Computer Science and Engineering PES University,
Bengaluru 560003, India
anurag.khanra306@gmail.com, badriprasad@pes.edu

Abstract. With the increase in the usage of the internet for gaining and providing information and knowledge, questionnaire forums are becoming popular means of the same for both teenagers and adults. Users post questions and mark them with the topic that they are related to (known as tags). Post this, users always expect quick answers/solutions from known and reliable sources. However, a large number of posted questions remain unanswered due to erroneous and huge number of tags. In our system, we propose an automated method to generate these tags using machine learning and deep learning techniques. This mainly helps in standardizing the tag content for similar questions and the total number of tags that the system needs to deal with. These tags can be generated from the content of the question provided by the user. In addition to this, for questionnaire forums related to programming questions like StackOverflow and StackExchange, our system aims to use the code snippets provided by the user to detect some common software vulnerabilities using deep learning techniques. Software vulnerabilities are weaknesses in software that may be exploited by malicious individuals. These vulnerabilities can arise from coding errors, flaws in design, or inadequate security measures, and can result in unauthorized access, data breaches, or system availability. Detection of these vulnerabilities beforehand will benefit the user and prevent any sort of security threats in their system.

Keywords: Question Tagging · Software Vulnerability Detection · Deep Learning · Convolutional Neural Networks · Natural Language Processing

1 Introduction

StackOverflow is an online forum where users can post questions related to programming, which are then directed to relevant experts for answers. However, StackOverflow does not verify the validity or security of the questions posed.

P. Das et al. (Eds.): AMRIT 2023, CCIS 1954, pp. 198–208, 2024.
https://doi.org/10.1007/978-3-031-47221-3_18

While most answers are generally relevant and useful, there is a risk that hidden issues may exist within the question itself. Although an incomplete solution may technically solve the problem at hand, such code can still contain vulnerabilities that may have devastating consequences for the user or application if left unpatched. For example, the first stack overflow solution suggested by Google for some of the most common vulnerabilities in C may be incomplete or inadequate, potentially causing problems for new developers relying on StackOverflow as a comprehensive resource.

To address this issue, we propose using deep learning techniques to detect potential vulnerabilities in code, which can then be tagged and categorized alongside existing tags generated from the user's question. This approach would provide users with warnings about possible mistakes or issues that may arise. We plan to train a deep learning model to identify common vulnerabilities in C and C++ (as these languages are particularly prone to certain types of vulnerabilities not seen in higher-level languages), and pass user-provided code from StackOverflow through this model to detect any such vulnerabilities.

Some of the most common C/C++ vulnerabilities we aim to detect include:

- CWE - 119: Improper Restriction of Operations within the Bounds of a Memory Buffer
- CWE - 120: Copying data from buffer without checking for its size. Eg: strcpy
- CWE - 469: Use of Pointer Subtraction to Determine Size
- CWE - 476: NULL Pointer Dereference

In online forums where users post questions and expect quick answers from reliable sources, tags are used to label the topics related to the questions. However, a significant number of posted questions remain unanswered due to the large number of erroneous and irrelevant tags used. Tags such as "object," "programming-challenge," or "beginner" provide no insight into the specific question or problem faced by the user. Additionally, duplicate tags, such as "python" and "python3," are used almost always in pairs, taking up one of the five slots available for tags and providing no additional insights.

Furthermore, the possibility of wrong tags is also considerable, as the user may not fully understand the exact issue with their code, leading to personal biases when adding tags to the question. To address these issues, machine learning techniques can be employed to generate tags automatically, removing the burden of assigning relevant and legitimate tags from the user. This approach can increase the accuracy and relevancy of tags, reducing the number of unanswered questions and improving the overall user experience in online forums.

2   Related Work

The objective of the research conducted by R. Russell et al. [18] is to develop an effective approach for detecting software vulnerabilities in code snippets written in C and C++ programming languages, using Machine Learning and Deep

Learning techniques. The authors compare their approach with alternate methods such as Static and Dynamic Analysers. To accomplish this task, the authors collected and curated data from various sources, including SATE IV Juliet Test Suite, GitHub, and Debian-based Linux distributions, which included a million functions of C and C++.

To prevent overfitting of the model due to the large and redundant code snippets, the authors used a custom lexer for source lexing. Afterward, the labelled data set was created by labelling the functions collected into the vulnerabilities they contained. The authors used various methods like static analysers, dynamic analysers and commit message tagging to label the dataset.

The authors employed two methods for vulnerability detection. Firstly, they used Neural network classification and representation learning, where they used a word-to-vec-based embedding with $k = 13$, followed by CNN and RNN for feature extraction. A fully connected classifier was used after feature extraction. Secondly, they used Ensemble learning on neural representations, where the output of CNN and RNN layers was passed to ensemble classifiers such as Random Forest.

The results demonstrated that CNN models outperformed RNN models with higher recall, and Ensemble classifiers performed better than Neural Networks. However, the authors acknowledged the possibility of further improvements, such as improving labels, detecting style violations, commit categorization, and algorithm/task classification. Overall, the research conducted by R. Russell et al. [18] provides a promising approach for detecting software vulnerabilities in code snippets of C and C++ programming languages using Machine Learning and Deep Learning techniques.

Cedeño González et al. [3] proposed a novel approach to generate tags for questions using a multi-class classification technique. They compared their approach with two other major methods in the field, namely TF-IDF and discriminant classifiers, using a dataset collected from the StackOverflow website. The authors employed traditional NLP pre-processing techniques and a Computer Science related lexicon to preserve specific terms. They also used a Linear Support Vector Classifier (LSVC) feature selection method with the L1 norm as a penalty function.

To prepare the data, each tag in the training set was assigned a position, and a separate classifier was used for each tag/position. The authors used two classification models, SVM and Naive Bayes, to evaluate the performance of their approach. Overall, they observed that the SVC classifier performed better than the NB classifier. Notably, the feature selection method had a minimal impact on the classifier performance.

This study makes a valuable contribution to the field of question tagging and demonstrates the effectiveness of a multi-class classification approach using LSVC feature selection. However, future research can investigate the use of additional features, such as improved labels, to enhance the performance of the classifier. Additionally, exploring the use of deep learning models for question tagging could be beneficial.

J. Dietrich et al. [6] investigated the relationship between tags and posts on Stack Overflow, specifically examining whether tags were associated with posts and whether they were related to programming language names. The authors analyzed the SOTorrent dataset to identify code snippets with tags, finding that 48,807,762 code snippets had at least one associated tag. The top 20 tags were determined using the TIOBE index.

To implement the project, the authors loaded the SOTorrent data into a MySQL dataset and created two tables: Snippets with tags from SOTorrent and Tags generated using Linguist. They used Java and R scripts to process the intermediate data. However, they noted limitations in tag normalization accuracy and representation of all programming languages. Therefore, the authors cautioned against relying on a single strategy, emphasizing the need for a mix of systems.

In their work, T. Saini and S. Tripathi [19] argue that the identification of relevant tags in a forum is critical to reduce the search space for duplicate questions and to group similar topics for easier browsing. To achieve this, they implemented a multilabel classification using a one-vs-all approach with SVM and linear kernels. They also explored two other approaches, the Naive Bayes method and Feature Extraction method, to improve the performance of the model on different sections of the dataset. Additionally, they employed lexical and syntactical analysis techniques such as parsing and tokenizing to enhance the efficiency of the model.

The dataset used in this study consisted of a large corpus of approximately 6 million questions from a forum. Each entry in the dataset included a question id, the question title, and the question body, which comprised the user's question in natural language and/or code snippets if available.

The authors used various techniques to analyze the dataset, including "Bayes Analysis on Indexed Dump," which used the classical naive-bayes method to generate tags using only the question title. They also used "SVM Analysis of List of Titles" and "Unique Feature Extraction Approach," in which the given question was converted into a vector, and the alignment of the vector with the labels was used to predict the tags. The results of the study indicated that, in general, SVM outperformed Bernoulli and Multinomial Naive Bayes, with performance improvements ranging from 30 to 50%.

In recent times, detecting software vulnerabilities has primarily relied on either analysing the source code via static methods or executing the final executable. However, researchers X. Zhang, L. Shao, and J. Zheng [22] have proposed alternative methods to detect vulnerabilities in software. One such method is fuzzing, where a series of semi-valid test data is automatically passed to the executable program to check for undefined or incorrect behavior, which may result in program crashes, depending on the software's reliability. Another proposed method is IDA-based assembling code audit, which involves reverse engineering and decomposing the executable to check for vulnerabilities. Few of the methods to generate the same are:

- This approach includes using buffer overflow vulnerabilities by overloading the input or buffer with a large string that exceeds the storage's boundaries, among others.
- Integer overflow and underflow vulnerabilities can also be detected by passing border values that can cause unintended effects such as integer wrap-around. For instance, values such as "−1, 0, 1, 0xff, 0xffff and 0xffffffff" can be used for this purpose.

However, most known vulnerabilities require specific constraints to be triggered, such as a particular input or data, without which the program may crash. Therefore, detecting the vulnerability's presence can be challenging when analysing the output or response. To overcome this challenge, one proposed approach is to use stack trace, etc., to track the functions and procedural calls performed by the software under consideration effectively. This method may not be affected by anti-debugging tools commonly used by proprietary software to prevent their code from being decompiled by competitors.

3   Proposed Solution

3.1   Question Tagging

Data. This dataset contains data to be used for the Question Tag Prediction Model. This is organized into 3 tables:

- **Questions** contains the text posted by the users, used for creation of question tags, and expert predection. Also contains the metadata of when the question was posted and at what time, the stack-overflow user ID.
- **Answers** containing the body, creation date, score, and owner ID.
- **Tags** containing the tags.

Pre-processing. The data for Question tagging were converted to dataframes and these data frames were joined using an "ID" field. Irrelevant columns in the resulting data frame were dropped. Questions with score less than 5 were not properly framed or were not tagged appropriately and hence were not used further. Based on the tags occurrence count, the questions with the top 50 common tags were retained for further analysis. Since the dataset contained questions scraped from StackOverflow, the question title and body consisted of non ASCII characters, HTML tags, email-ids and URLs. These were removed using a pre-processing library. For the next steps of text pre-processing, the NLTK library was used for lemmatization and stop word removal. Lemmatization in NLP is a form of grouping together the words of a language with minor inflections, so that the model can treat these similar words as the same to reduce complexity, to prevent overfitting, etc. Stopwords are words such as "and", "the", "if", etc. which do not add any significant meaning while tokenization. The data was finally converted to One Hot Encoding with top 50 tags. Tokenization was done using a standard keras tokenizer with a threshold for maximum number of words as 5000 and maximum length of each question as 500.

Convolutional Neural Network. A single embedding layer was used for consuming the input. This was followed by 3 pairs of convolutional and pooling layers. Each convolutional layer had different dropout values to prevent overfitting with a Rectified Linear Unit activation function and a filter length of 300. This was followed by 2 dense layers of size 64 and 16 respectively with a Rectified Linear Unit activation function. The final dense layer was used as the output layer with a sigmoid activation function. The loss function used is binary cross entropy with an Adam optimizer. A batch size of 32 was used for training with 30 epochs. In an attempt to improve results and follow some of the state-of-the-art models, a static analysis method was ensembled with the current Convolutional Neural Network. The static analysis method uses a correlation matrix between the tags to predict which tags are related to other tags very closely and hence occur together very frequently. These highly correlated tags were combined with the originally predicted tags by the model and given as the final output. This method of ensembling static analysis along with the model prediction improved the recall of the overall model.

3.2   Software Vulnerability Detection

Data. The dataset is of the type HDF5(The Hierarchical Data Format version 5) which contains string snippets of code along with the top 4 vulnerabilities which are one hot encoded to show the presence of that vulnerability.

Pre-processing. The available data in HDF5 format is decoded from binary strings to UTF-8 format. Subsequently, the code is cleaned by removing unnecessary components such as comments and variable names, as these may introduce bias. Single-line comments are removed using regex, and each token is compared to a list of all C/C++ main library keywords. If a token is not present, it is replaced with the term "variable" to remap all variables to the same. White spaces, such as tabs and newlines, which do not affect the code's execution, are removed. Multi-line comments are also eliminated. The resulting pre-processed text is then ready for training. Following this, the Boolean vulnerabilities are converted to integer values, similar to One Hot Encoding. As evidenced by the token count, all remaining tokens are pertinent to the code.

Convolutional Neural Network. The present study utilized a CNN model for predicting the presence of vulnerabilities in code snippets. The model was found to perform best for the given use case. The architecture of the final CNN model comprised an input layer for tokenized strings of code snippets, followed by 2 Convolution layers on 1 Dimension with different dropout rates, adjusted via tuning. This was followed by 3 hidden dense layers, each using the "ReLU" activation function. The final output layer produced 5 floating point numbers as output, determining the probability of vulnerability existence in the code snippet. A threshold was estimated after fine-tuning the model, converting the floating point output to Boolean output. The Adam Optimizer was employed for its benefits such as fast training, simple implementation, and verifiable results.

Let X be the input tensor of shape (batchsize, sequencelength, embeddingsize), and F be the set of learnable filters of shape (filterlength, embeddingsize, numfilters). The Conv1D layer applies the convolution operation with stride 1 and valid padding. The output feature map Y is computed as follows:

$$Y(i,j,k) = \sum_{m,l}[X(i+m,j,l) \cdot F(m,l,k)] + B(k) \tag{1}$$

Here, (i, j) represents the spatial coordinates of the output feature map, k represents the index of the filter, (m, l) represents the spatial coordinates within the filter, and B(k) denotes the bias term for the k-th filter.

The model was trained over 30 epochs, using a 70:30 split between training and testing, with 20% being reserved for validation during training.

3.3   Results

Question Tagging and Vulnerability Detection problems are both multi-class classification problems. Hence, metrics such as precision, recall and F1 score were used to determine their performance (Tables 1 and 2).

Table 1. Comparison with state of the art models (Question Tagging)

Model	Recall@5	Precision@5
SOTagRec	0.817	0.343
TagCombine	0.595	0.221
TagMulRec	0.680	0.284
EnTagRec	0.805	0.346
CNN(Proposed model)	0.844	0.239

Table 2. Comparision with state of the art models(Vulnerability Detection

Paper/Method	Recall	F1 score
Automated Vulnerability Detection in Source Code using Deep Representation Learning	Private test data used, hence unable to compare	0.56
Static Tests (Clang/cppcheck)	<0.50	<0.25
CNN (our model)	0.55	0.78

Inferences. A few inferences observed after obtaining and analyzing the results are as follows:

- Tags like "database", "performance" and "algorithm" have a relatively lower precision than the other tags. This is probably because there is no key distinguishing feature in questions with these tags and these topics are very generalized.
- Tags like "python", "JavaScript" and "PHP" (programming languages) have a relatively higher precision than the other tags.
- A usual trend observed in both, question tagging and vulnerability detection is that an increase in the number of training samples for a particular tag/vulnerability increases the precision and recall for that particular tag/vulnerability.
- The data cleaning steps are critical in improving results and decreasing training time.

Drawbacks. The approach taken and models built have the following drawbacks/failure cases:

- The number of tags generated are lesser than expected for actual questions. This is because most questions do not revolve around many major topics and the model is trained for only the top 50 tags.
- Most of the code snippets in the training data for vulnerability detection are not vulnerable which gives rise to problems while dealing with a skewed data set.
- Samples of code snippets with CWE-469 are low in number causing the recall to decrease for this vulnerability.

4   Conclusion

This research project aimed to enhance the user experience for questionnaire forums, such as StackOverflow, by automating the generation of relevant tags and the detection of potential software vulnerabilities for C and C++ languages, using deep learning techniques. The project yielded promising results; however, there is still room for improvement in the current approach. Future work can be focused on incorporating additional relevant features to further enhance the accuracy and effectiveness of the system.

5   Future Work

The future scope of the project involves several areas of potential improvement and expansion.

Improving Accuracy of Vulnerability Detection. While the current model has shown promising results, there is still scope for improvement in terms of accurately detecting vulnerabilities. To enhance the accuracy of the model, increasing the dataset size is essential. By gathering a larger and more diverse collection of data points, the CNN model can learn from a wider range of examples, leading to improved accuracy and better generalization. Experimenting with different CNN architectures, such as increasing the depth or width of the network or utilizing more advanced models like ResNet or DenseNet, enables the model to capture more intricate patterns and features within the vulnerability data. Lastly, by systematically adjusting hyperparameters like learning rate, batch size, optimizer, regularization techniques, and activation functions, the model can find the optimal configuration that maximizes accuracy.

Expanding the Scope of the Project. The project can be extended beyond just questionnaire forums like StackOverflow to other platforms like GitHub and other code repositories. This can help developers identify potential vulnerabilities in their code before it is released to the public.

Larger Number of Tags for Training the Model. This could help improve the accuracy of tag prediction and make the system more useful for users of the forum. Additionally, incorporating a larger number of vulnerabilities into the scope of the project could enhance the model's ability to detect potential software security issues in code snippets.

Incorporating More Programming Languages. Currently, the project only focuses on C and C++, but it can be extended to include other languages like Java, Python, and more. This would require training the model on new datasets and adjusting the architecture of the model to suit the requirements of each language.

Detection of Algorithm. Another area of potential expansion is the detection of algorithm/description of code snippets. This could help users better understand the purpose and functionality of the code they are working with, as well as identify potential areas for improvement or optimization. Incorporating natural language processing techniques could be helpful in achieving this goal.

References

1. Alayba, A.M., Palade, V., England, M., Iqbal, R.: A combined CNN and LSTM model for Arabic sentiment analysis. In: Holzinger, A., Kieseberg, P., Tjoa, A.M., Weippl, E. (eds.) CD-MAKE 2018. LNCS, vol. 11015, pp. 179–191. Springer, Cham (2018). https://doi.org/10.1007/978-3-319-99740-7_12

2. Amin, R., Sworna, N.S., Hossain, N.: Multiclass classification for Bangla news tags with parallel CNN using word level data augmentation. In: 2020 IEEE Region 10 Symposium (TENSYMP), pp. 174–177 (2020). https://doi.org/10.1109/TENSYMP50017.2020.9230981

3. Cedeño González, J.R., Flores Romero, J.J., Guerrero, M.G., Calderón, F.: Multi-class multi-tag classifier system for stackoverflow questions. In: 2015 IEEE International Autumn Meeting on Power, Electronics and Computing (ROPEC), pp. 1–6 (2015). https://doi.org/10.1109/ROPEC.2015.7395121

4. Chauhan, V.K., Kumar, A.: Vulnerability detection in source code using deep representation learning. In: 2022 11th International Conference on System Modeling & Advancement in Research Trends (SMART), pp. 1515–1520 (2022). https://doi.org/10.1109/SMART55829.2022.10047614

5. Chernis, B., Verma, R.: Machine learning methods for software vulnerability detection. In: Proceedings of the Fourth ACM International Workshop on Security and Privacy Analytics, New York, NY, USA, pp. 31–39. Association for Computing Machinery (2018). https://doi.org/10.1145/3180445.3180453

6. Dietrich, J., Luczak-Roesch, M., Dalefield, E.: Man vs machine - a study into language identification of stack overflow code snippets. In: 2019 IEEE/ACM 16th International Conference on Mining Software Repositories (MSR), pp. 205–209 (2019). https://doi.org/10.1109/MSR.2019.00041

7. Farooq, A., Anwar, S., Awais, M., Rehman, S.: A deep CNN based multi-class classification of Alzheimer's disease using MRI. In: 2017 IEEE International Conference on Imaging Systems and Techniques (IST), pp. 1–6 (2017). https://doi.org/10.1109/IST.2017.8261460

8. Hong, J., Fang, M.: Keyword extraction and semantic tag prediction (2013). unpublished http://cs229.stanford.edu/proj2013/FangHong-Keyword%20Extraction%20and%20Semantic%20Tag%20Prediction.pdf

9. Jain, V., Lodhavia, J.: Automatic question tagging using k-nearest neighbors and random forest. In: 2020 International Conference on Intelligent Systems and Computer Vision (ISCV), pp. 1–4 (06 2020). https://doi.org/10.1109/ISCV49265.2020.9204309

10. Li, N., Liu, Y., Li, L., Wang, Y.: Smart contract vulnerability detection based on deep and cross network. In: 2022 3rd International Conference on Computer Vision, Image and Deep Learning International Conference on Computer Engineering and Applications (CVIDL & ICCEA), pp. 533–536 (2022). https://doi.org/10.1109/CVIDLICCEA56201.2022.9824581

11. Lin, Y.H., Chen, H.H.: Tag propagation and cost-sensitive learning for music auto-tagging. IEEE Trans. Multimedia 23, 1605–1616 (2021). https://doi.org/10.1109/TMM.2020.3001521

12. Mahyari, A.: A hierarchical deep neural network for detecting lines of codes with vulnerabilities. In: 2022 IEEE 22nd International Conference on Software Quality, Reliability, and Security Companion (QRS-C), pp. 1–7 (2022). https://doi.org/10.1109/QRS-C57518.2022.00011

13. Nishizono, K., Morisakl, S., Vivanco, R., Matsumoto, K.: Source code comprehension strategies and metrics to predict comprehension effort in software maintenance and evolution tasks - an empirical study with industry practitioners. In: 2011 27th IEEE International Conference on Software Maintenance (ICSM), pp. 473–481 (2011). https://doi.org/10.1109/ICSM.2011.6080814

14. Park, S., Song, J.H., Kim, Y.: A neural language model for multi-dimensional textual data based on cnn-lstm network. In: 2018 19th IEEE/ACIS International Conference on Software Engineering, Artificial Intelligence, Networking and Parallel/Distributed Computing (SNPD), pp. 212–217 (2018). https://doi.org/10.1109/SNPD.2018.8441130

15. Radenović, F., Tolias, G., Chum, O.: CNN image retrieval learns from BoW: unsupervised fine-tuning with hard examples. In: Leibe, B., Matas, J., Sebe, N., Welling, M. (eds.) ECCV 2016, Part I. LNCS, vol. 9905, pp. 3–20. Springer, Cham (2016). https://doi.org/10.1007/978-3-319-46448-0_1

16. Ramesh, V., Abraham, S., Vinod, P., Mohamed, I., Visaggio, C.A., Laudanna, S.: Automatic classification of vulnerabilities using deep learning and machine learning algorithms. In: 2021 International Joint Conference on Neural Networks (IJCNN), pp. 1–8 (2021). https://doi.org/10.1109/IJCNN52387.2021.9534259

17. Ruggahakotuwa, L., Rupasinghe, L., Abeygunawardhana, P.: Code vulnerability identification and code improvement using advanced machine learning. In: 2019 International Conference on Advancements in Computing (ICAC), pp. 186–191 (2019). https://doi.org/10.1109/ICAC49085.2019.9103400

18. Russell, R.L., et al.: Automated vulnerability detection in source code using deep representation learning. In: 2018 17th IEEE International Conference on Machine Learning and Applications (ICMLA), pp. 1–6 (2018). https://doi.org/10.1109/ICMLA.2018.00120

19. Saini, T., Tripathi, S.: Predicting tags for stack overflow questions using different classifiers. In: 2018 4th International Conference on Recent Advances in Information Technology (RAIT), pp. 1–5 (2018). https://doi.org/10.1109/RAIT.2018.8389059

20. Wang, T., Wei, T., Gu, G., Zou, W.: Taintscope: a checksum-aware directed fuzzing tool for automatic software vulnerability detection. In: 2010 IEEE Symposium on Security and Privacy, pp. 497–512. IEEE (2010)

21. Yamaguchi, F.: Pattern-based methods for vulnerability discovery. It-Inf. Technol. 59(2), 101–106 (2017)

22. Zhang, X.S., Shao, L., Zheng, J.: A novel method of software vulnerability detection based on fuzzing technique. In: 2008 International Conference on Apperceiving Computing and Intelligence Analysis, pp. 270–273 (2008). https://doi.org/10.1109/ICACIA.2008.4770021

23. Zoubi, Q., Alsmadi, I., Abul-Huda, B.: Study the impact of improving source code on software metrics. In: 2012 International Conference on Computer, Information and Telecommunication Systems (CITS), pp. 1–5. IEEE (2012)

A Comprehensive Study of the Performances of Imbalanced Data Learning Methods with Different Optimization Techniques

Debashis Roy[1]($\boxtimes$), Utathya Aich[2], Anandarup Roy[3], and Utpal Roy[1]

[1] Department of Computer and System Sciences, Visva Bharati University, Santiniketan, India
debashisjcc@gmail.com, utpal.roy@visva-bharati.ac.in
[2] PGET, CNH Industrial, Gurgaon, Gurugram, India
[3] Sarojini Naidu College for Women, Dum Dum, Kolkata 700028, India
roy.anandarup@sncwgs.ac.in

Abstract. Imbalanced classification is most important problem in machine learning, where the distribution of classes in the training dataset is skewed towards one class. So, insufficient representation of data in interested class (minority class) leads to misclassification by the canonical classifier, where learning by training datasets is biased towards the majority class. Several important applications are imbalanced in nature. Many different methods were proposed by the researchers to improve classification performance in an imbalanced dataset. The most common methods are resampling, cost-sensitive learning, ensemble methods, anomaly detection, etc. But there is a huge research gap in the prediction of an imbalanced or highly imbalanced dataset because the proposed methods are not assured about prediction accuracy. In cases of imbalanced datasets, hyperparameter tuning becomes more important to improve accuracy because a poorly tuned predictive model can easily neglect the interested classes (rare events), resulting in poor performance. This paper focused on comparative studies of the performance of different optimization methods with a proposed classifier for an imbalanced dataset.

Keywords: Big Data · Machine learning · Imbalanced dataset · Hyper-parameter tuning · Brain Stroke prediction · Customer Churn Prediction

1 Introduction

The problem of classification occurs when the number of instances in one class (the minority class) is much smaller than the number of instances in the other class (the majority class) [1]. Even it become highly imbalanced if ratio between majority and minority class is like 10000:10. This is a common phenomenon for many real world applications, such as drug effect and risk prediction [2], early stage cancer prediction [3], fraud detection [4], churn prediction [5], disease prediction [6, 7], object recognition [8], business management [9] etc. Because the majority class dominates the training dataset, the predictive model fails to detect an interested class or a rare event [10]. A degree of

P. Das et al. (Eds.): AMRIT 2023, CCIS 1954, pp. 209–228, 2024.
https://doi.org/10.1007/978-3-031-47221-3_19

imbalance can be measured by an imbalanced ratio (IMB ratio = majority class/minority class). However, while the accuracy of such models can be high, they are completely biased in favour of the majority [11]. This can result in serious consequences, such as missing important information and making incorrect predictions. There are several methods to handle imbalanced dataset, which fall into three major categories, data-level methods, special-purpose learning methods, and hybrid methods. However, both data-level and algorithm-level methods face some difficulties [12].

So, unbalanced dataset optimization with efficient resampling techniques is the most challenging work, and in the same way, an efficient algorithm that is balanced and may work under dataset shift conditions. Another efficient approach for the classification of an imbalanced dataset is hyperparameter tuning with class weight optimization [13]. Once you have chosen a suitable algorithm and a good evaluation metric, for the imbalanced dataset, you can start tuning the hyperparameters. Generally, naive optimization algorithms are used to tune hyperparameters, such as grid search and randomised search. Another two approaches are stochastic optimization algorithms like the stochastic hill climbing algorithm [14] and genetic optimization algorithms [15, 16] for hyperparameter tuning for the classification of imbalanced datasets. Researchers proposed different optimization algorithms for hyperparameter tuning with regard to the learning and validation of an imbalanced dataset.

We conducted two experiments in this paper: the first used canonical algorithms KNN [18], SVM [19], and Logistic Regression in conjunction with SMOTE-Tomeklink [21], and the second used ensemble classifiers XGBoost [20], LightGBM, CatBoost along with decision tree [17], and random forest [20]. In the first experiment, we worked with highly imbalanced cerebral stroke data, and for the second one, we used churn prediction data. In both cases, we used hyperparameter optimization as Grid search, Random search, Tree of Parzen estimator (TPE), Genetic Algorithm. Here we study a comparative analysis of the performance of the above optimization algorithms on an imbalanced dataset.

2   Strategy for Imbalanced Data Classification

2.1   Performance Metrics

Metric selection plays a vital role in imbalanced learning. The selection of the wrong metric can mean choosing an improper classification model [22, 23]. Metric selection can be influenced by your application, algorithm choice, and the nature of the dataset. For example, accuracy is a popular performance metric for machine learning, but it can be misleading and biased towards the majority class in cases of imbalanced datasets. Guido et al. (2020) [24] tested G-Mean against accuracy on two imbalanced datasets by optimizing the hyperparameter of a support vector machine using a genetic algorithm and concluded that G-Mean is more appropriate for evaluating model performance than accuracy.

Commonly used metrics for classification and hyperparameter tuning of imbalanced datasets are balanced accuracy [25], G-mean [26], AUC, and F1-score [27]. Zhang et al. (2022) [28] used 7 metrics in addition to PPV12 and NPV12 to improve the classifier's performance with hyperparameter tuning for imbalanced Alzheimer's Disease data. Burduk [29] proposed a new metric, HMNC, for imbalanced classification in two ways: one when the majority class is larger than the imbalance ratio (IR) and another when the majority class is smaller than the IR, though the author proposed a different formula to calculate the IR. As a result, proper metric selection, along with hyperparameter tuning, is critical for an efficient classification result with an imbalanced dataset.

2.2 Data Level Methods

There are three types of data level approaches used to handle an imbalanced dataset: oversampling, undersampling, and hybrid methods.

- The process of introducing new synthesized data into the minority class is known as oversampling. New data is generated from existing data. There are many proposed oversampling methods, and the most popularly used is SMOTE [30] and different varieties of SMOTE such as borderline-SMOTE [31], modified SMOTE (MSMOTE) [32], kernel-based SMOTE [33], ADASYN [34], etc.
- To balance the dataset, undersampling removes data from the majority class. Edited nearest neighbours [35] and Tomek links [36] are two popular undersampling methods.
- The hybrid method is a combination of oversampling and undersampling and sometimes works more efficiently for some applications. Some newly proposed hybrid methods are SMOTE + PSO + C5 [26], Borderline-SMOTE SVM [37], RUS-IPF [38], etc.

Rekha et al. (2019) [39] worked with 15 different datasets and showed that the oversampling technique takes longer training time and is less efficient (in terms of memory, due to the increased number of training instances) than the undersampling technique, and that it suffers from high computational costs (for pre-processing the data). Vargas et al. (2023) [40] reviewed 9927 papers on sampling techniques and concluded that the best performing and most popular techniques for the future are cluster-based sampling, oversampling, and hybrid sampling. There are conflicts and research gaps in data level methods because data distribution should be optimal according to user goals.

2.3 Algorithm-Level Approach

Cost-sensitive learning frameworks lie in between data- and algorithm-level approaches. Such methods assign different costs to instances, modify the learning algorithm to incorporate varying penalties, and accept the cost. Well-known methods in this category are cost sensitive SVM [41] and stochastic gradient boosting [42]. There are many cost-sensitive algorithms classified as linear, nonlinear, and ensemble algorithms. Some ensemble algorithms are AdaBoost [43], XgBoost [44], BABoost [45], etc.

2.4 One Class Classifier

This algorithm focuses on a target group to create a data description. This algorithm is used for outlier detection and anomaly detection at the time of classification tasks. Some one-class algorithms are one-class support vector machines [46], isolation forests, local outlier factors [47] etc. (Table 1).

Table 1. Some challenges in data level and algorithm level strategies.

Challenges	Data-level	Algorithm level
1	The amount of data added to the minority class or removed from the majority class is defined as optimal data distribution	Create an algorithm that works with data shifts and focuses on the minority class
2	Ratio of data resampling and distribution according to user preference	An algorithm that is optimal and tailored to the user's goals
3	Effective optimal resampling method based on hyper-parameter optimization, suitable for all situations	Algorithm and optimization method to design adequate algorithms for all scenarios
4	Overfitting for additional data from the minority class and removal of important data for the undersampling method	Design algorithms that efficiently work for the classification of imbalanced datasets along with overlapping and other data complexities

2.5 Challenges on Handling Class Imbalance

3 Hyper-parameter Tuning

Every model has some default parameters that are learned from the training dataset and updated at learning time to predict, classify, etc. We could not control the model parameter because it was derived from data. The hyperparameters, on the other hand, specify structural information about the type of model itself, such as whether or not we are using classification or linear regression, the ideal architecture for a neural network, the number of layers, the type of filters, regularisation strength, and batch size etc. They are not learn; they are defined before training. Where as hyperparameter space is a range of values that is being explored to tune the hyperparameters. Finding the configuration of hyperparameters that produces the best performance is referred to as "hyperparameter tuning" or "hyperparameter optimization." The main task now is to find the best hyperparameters in the search space for a given model.

Hutter et al. (2014) [48] propose a more comprehensive approach for evaluating the significance of individual hyperparameters. Similarly Probst et al. (2019) [49] showed the importance of hyperparameters to get efficient results from a model. They proposed the best hyperparameter configuration for a given dataset.

Hyperparameters of the Ml model can be continuous (learning rate, coefficient, weight balanced factor, etc.) or discrete (max_depth, number_of_hidden_layers, batch size etc.). The problem of risk/error minimization is frequently relaxed by continuous optimization. Many parameterized models [50] involve continuous optimization in solving learning problems. Where as inference problems within structured spaces, specific learning problems, and auxiliary problems like feature selection, data subset selection, data summarization, neural architecture search, etc. all include discrete optimization. The architecture of the model influences the selection of critical hyperparameters, such as the level of polynomial features that should be used in a linear model. What number of neurons should we preserve in each layer of the neural network? In a random forest, how many trees are needed? learning rate of the gradient descent method etc.

Yang et al. (2020) [51] described excellently several optimization methods and how to implement them to machine learning models. Likewise Bischl et al. (2023) [52] demonstrated important hyperparameter choices and how to combine them with pipelines for parallelization and computational time reduction.

There are numerous methods for hyperparameter tuning or hyperparameter optimization such as grid search, random search [53], successive halving [54], Bayesian Optimization [55], Gradient Based [56], Hyperband [57], Genetic Algorithm [58] etc.

For machine learning, there are a number of hyperparameter optimization libraries that can help automate and accelerate the process of finding the optimal collection of hyperparameters. These are a few well-known hyperparameter optimization libraries such as Scikit-Optimize, Hyperopt, Optuna, Osprey, TPOT, Ray Tune, SMAC, Keras Tuner etc.

Shekhar et al. (2021) [59] described a comparative study of hyperparameter optimization tools using two benchmarks and showed that Optuna performed better for the CASH problem and HyperOpt for the MLP problem.

4 How Does Hyperparameter Optimization Improve the Classification Result for an Imbalanced Dataset?

4.1 Data Level

Another important decision is how much data will be added to the minority class in oversampling or removed from the majority class in undersampling techniques. Here it is necessary to design efficient resampling methods using hyperparameter tuning that produce an optimal distribution of balanced data.

Pelin Akin (2023) [15] showed that class ratios and k parameters can be optimized by genetic algorithm to boost the SMOTE algorithm performance.

Vaseekaran (2022) [60] give the idea to identify the efficient resampling strategy using hyperparameter tuning along with pipeline class library of scikit-learn's. This mechanism can work potentially effective if researcher implements it as future aspect with classifiers.

Sun et al. (2021) [61] proposed an auto-sampling technique which based on optimization sampling schedules and it able worked for high dimensional hyperparameters.

Li et al. [62] resampled the biologically imbalanced data using a Bat-inspired algorithm and a PSO optimization algorithm, depending on the properties of the biological datasets.

Saglam et al. (2021) [63] demonstrated undersampling methods using bee colony and partial herd evolution optimization methods.

The direction of hyperparameter optimization of resampling techniques and their advantages were demonstrated by Moniz et al. (2021) [64].

Random search and Tree Parzen estimators based on Bayesian optimization on resampling methods and classification were carried out by Nguyen et al. (2021) [65]. They explore with a search space consisting of 64 pertinent hyperparameters, 21 resampling strategies, and 4 classification algorithms.

Using hyperparameter tuning, optimal data distribution that is adequate and in accordance with the user's goal could be a future scope.

4.2 Algorithm Level

Many researchers in different applications utilized and acquired the best results using hyperparameter tuning with classifiers for imbalanced or highly imbalanced datasets.

Hancock et al. (2021) [66] demonstrated in an experiment that classifiers with hyperparameter tuning always outperformed their default values.

Zhang et al. (2022) [28] demonstrated that SVM with a hyperparameter tuning model can produce the best results and perform well with imbalanced Alzheimer's disease data.

Other work applied hyperparameter optimization along with cost-sensitive SVM using genetic algorithms. Guido et al. (2022) [16] demonstrated effective classification results with an imbalanced dataset using a genetic optimization algorithm versus the grid search method.

Similarly Sharma et al. (2021) [67] used RAO optimization algorithms to classify Parkinson's disease, and they also compared the performance with other methods. So, both classification algorithms and resampling techniques involve some hyperparameters that can be tuned. Here we describe a few recent works in Table 2.

Table 2. An overview of some recent works regarding hyperparameter tuning with an imbalanced dataset.

Authors	Hyperparameter Optimization Algorithm	Data Set	Description
Panda et al. (2023) [68]	Genetic Optimization Algo.	Imbalanced data of urban and rural energy consumers	Unbalanced classification of urban and rural energy consumption was resolved using a Monte Carlo undersampling and genetic optimization approach for each classifier

(continued)

Table 2. (*continued*)

Authors	Hyperparameter Optimization Algorithm	Data Set	Description
Bertsimas et al. [69]	Robust Optimization (RO) Algo.	20 real world imbalanced datasets	Used robust optimization to train logistic regression and SVM on imbalanced datasets
Rosales-Pérez et al. (2022) [70]	Evolutionary Bilevel Optimization	70 imbalanced datasets	They proposed a bilevel cost-sensitive SVM (EBCS-SVM), which handles imbalanced datasets with trained support vectors with hyperparameter optimization and misclassification costs
Nishat et al. (2022) [71]	Grid Search and Random Search Optimization	Imbalanced heart failure dataset	Investigate an experiment with an imbalanced heart failure dataset using 6 classifiers and SMOTE-ENN with grid search and random search optimization techniques
Zhang et al. (2022) [28]	Grid Search Optimization Algo.	Alzheimer's disease dataset	They applied a multicore high performance workflow to run parallel SVM hyperparameter tuning to learn alzheimer's disease
Tharwat et al. (2019) [14]	Ski-driver (SSD) optimization algorithm	8 imbalanced dataset	They proposed the ski-driver (SSD) optimization algorithm along with SVM (SSD-SVM) on eight imbalanced datasets and compared it with grid search (PSO)

4.3 How to Use Hyperparameter Optimization to Handle Unbalanced Dataset Along with Data Complexities

The difficulty of classifying imbalanced data becomes increasingly challenging when data complexities are present. Here, overlap is significantly affecting how well imbalanced data are classified. Even distinguishing a class from an overlapped region is an open issue, but many researchers try to minimise the problem with other techniques.

Vuttipittayamongkol et al. [72] done an experiment on synthetic datasets to show how class overlap plays a major role for accuracy of minority class for imbalance datasets.

Kong et al. [73] use hyperparameter optimization to try to reduce the effect of overlap.

Hence, handling overlapped data and other complexities for an unbalanced dataset using hyperparameter adjustment may be a future effort.

5 Experiment

5.1 Experiment 1

In this study, we used **highly imbalanced datasets** (cerebral stroke) from the Kaggle database to run an experiment with an imbalanced ratio of **0.0168**, where negative class (majority class) is 32613 and interest class (minority class) is 550.

Here we used three classifiers: k-nearest neighbours (**KNN**), Logistic Regression, **SGD**-linear classifiers (SVM, **logistic regression**, etc.) with stochastic gradient descent (SGD) training, and **SMOTE-Tomek** to resample the dataset. As a hyperparameter optimization algorithm, we used **Grid Search** (Gridsearchcv), **Random Search** (Randomsearchcv), the Tree of Parzen estimator (**TPE**) using Hyperopt, and TPOT using a **genetic algorithm**. Use accuracy, ROC-AUC, F1-Score, and G-Mean as performance measurements.

Result

The experimental results are shown in Table 3, where we show the parameters that have been set using hyperparameter optimization techniques and the accuracy, ROC-AUC, F1-Score and G-Mean of different classifiers after and before hyperparameter tuning.

Table 3. The effectiveness of classifiers prior to and following hyperparameter adjustment.

Name	parameters	Accuracy	Roc_auc	F1score	Gmean
Logistic regression	Default	0.7572	0.6007	0.0835	0.7553
Grid_log_reg	penalty = "I2", C = 0.01	0.9027	0.85	0.2127	0.8483
TPOT_LOG	C = 0.01, penalty = 12	0.9027	0.85	0.2127	0.8483
Rand_log_reg	penalty = "I2", C = 0.001	0.9214	0.8882	0.2641	0.8876

(continued)

Table 3. (*continued*)

Name	parameters	Accuracy	Roc_auc	F1score	Gmean
Hyper_opt_TPE_LOG	penalty = 12, C = 0.001, solver = lbfgs	0.9214	0.8882	0.2641	0.8876
TPOT_KNN	algorithm = kd_tree, n_neighbors = 2	0.8738	0.5555	0.056	0.4474
Grid_K_nn	n_neighbors = 3, algorithm = ball_tree	0.8145	0.6007	0.0633	0.5585
KNearest	Default	0.7989	0.6143	0.0651	0.5839
Rand_K_nn	n_neighbors = 4, algorithm = kd_tree	0.841	0.5962	0.0666	0.5398
Hyper_opt_TPE_KNN	n_neighbors = 4, algorithm = ball_tred	0.841	0.5962	0.0666	0.5398
TPOT_SGD	loss = hinge, penalty = elasticnet, alpha = 0.001	0.1292	0.5537	0.0363	0.3374
SGD_class	Default	0.951	0.5445	0.0773	0.346
Hyper_opt_TPE_SGD	loss = hinge, penalty = elasticnet, alpha = 0.01	0.7743	0.7704	0.1009	0.7704
Grid_SGD	pss = modified_huber, penalty = elasticnet, alpha = 0.1000	0.7729	0.7733	0.1012	0.7733
Rand_SGD	loss = log, penalty = elasticnet, alpha = 0.0100	0.7759	0.6007	0.1015	0.7712

The accuracy of KNN increased after all hyperparameter optimization, but ROC-AUC and G-Mean were not improved, however, F1-Score improved after random search and TPE.

In the case of SGD, accuracy not improved, but ROC-AUC, F1-Score, and G-mean significantly higher after grid search, random search, and TPE compared to the genetic algorithm.

The accuracy, ROC-AUC, and G-Mean of Logistic Regression (LR) are improved after all hyperparameter tuning methods, but the improvement of F1-Score is measurably high after all hyperparameter settings.

Here we demonstrated F1-Score in Fig. 1 and ROC-AUC in Fig. 2 for all classifiers along with hyperparameter optimization algorithms.

5.2 Experiment 2

In the second experiment, we used Kaggle's imbalanced churn datasets with an imbalanced ratio of **0.3612**, where churn data was 1869 (for the interest class) and customer

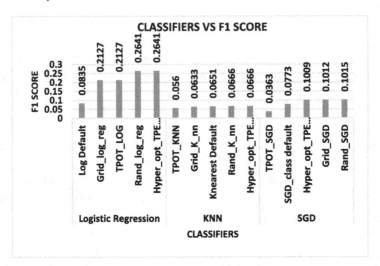

Fig. 1. F1-Score of all classifiers using methods with hyperparameter optimization

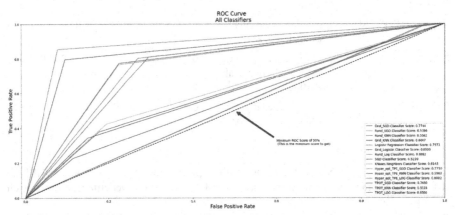

Fig. 2. Individual ROC-AUC growth before and after hyperparameter adjustment for each classifier

data was 5174 (for the majority class). Classifiers used include **Decision Tree**, **Random Forest**, **XGBoost**, **LightGBM** [74] and **CatBoost** [75]. However, hyperparameter optimization algorithms and performance metrics are the same as in experiment 1.

Result

The test results are displayed in Table 4, Here, we showcase the classifiers' performance reports prior to as well as after hyperparameter optimization settings.

Decision Tree: Accuracy and ROC-AUC of the decision tree are improved after all hyperparameter settings, where F1-score is improved after TPE, random search, and genetic optimization algorithms. G-Mean is increased after Random Search and Genetic optimization tuning.

Table 4. The performance of classifiers both before and after hyperparameter tweaking.

Name	parameters	Accuracy	Roc_auc	F1score	Gmean
Grid_DecisionTreeClassifier	{criterion: entropy, max_depth: 3, max_features: 7, min_samples_leaf: 1}	0.7768	0.6456	0.4654	0.5816
DecisionTreeClassifier	Default	0.7142	0.6371	0.4678	0.6154
Hyperopt_DecisionTreeClassifier	(criterion = entropy, max_depth = None, max_features = 7, min_samples_leaf = 2, random_state = 123)	0.7436	0.6457	0.4753	0.611
Rand_DecisionTreeClassifier	{criterion: gini, max_depth: None, max_features: 4, min_samples_leaf: 6}	0.764	0.6721	0.5174	0.6428
Tpot_DecisionTreeClassifier	(criterion:gini, max_depth:None, max_features:3, min_samples_leaf:7)	0.7725	0.6876	0.542	0.6632
RandomForestClassifier	Default	0.7839	0.6885	0.544	0.6577
Hyperopt_RandomForestClassifier	(bootstrap = True, max_depth = 90, max_features = 3, min_samples_leaf = 10, n_estimators = 100)	0.7967	0.6938	0.5536	0.6581
Tpot_RandomForestClassifier	(bootstrap = True, max_depth = 110, max_features = 2, min_samples_leaf = 4, min_samples_split = 10, n_estimators = 200)	0.8005	0.6958	0.5573	0.659
Rand_RandomForestClassifier	(bootstrap = True, max_depth = 50, max_features = auto, min_samples_leaf = 2, min_samples_split = 2, n_estimators = 1200)	0.7934	0.6967	0.5578	0.6654
Grid_RandomForestClassifier	(bootstrap = True, max_depth = 110, max_features = 3, min_samples_leaf = 3, min_samples_split = 8, n_estimators = 100)	0.8019	0.7036	0.57	0.6716

(*continued*)

Table 4. (*continued*)

Name	parameters	Accuracy	Roc_auc	F1score	Gmean
Tpot_XGBClassifier	(learning_rate = 0.1, max_depth = 4)	0.7967	0.7097	0.5782	0.685
Grid_XGBClassifier	(learning_rate = 0.20, max_depth = 8, min_child_weight = 5, gamma = 0.2, colsample_bytree = 0.5)	0.7768	0.6825	0.5341	0.6522
XGBClassifier	Default	0.7773	0.6942	0.5524	0.6712
Rand_XGBClassifier	(learning_rate = 0.20, max_depth = 8, min_child_weight = 5, gamma = 0.3, colsample_bytree = 0.5)	0.7839	0.6948	0.5538	0.6682
Hyperopt_XGBClassifier	(colsample_bytree = 0.7, gamma = 0.4, learning_rate = 0.2, max_depth = 12, min_child_weight = 5)	0.7829	0.6987	0.5596	0.6751
Hyperopt_CatBoostClassifier	(depth = 10, learning_rate = 0.01, iterations = 60)	0.7976	0.6905	0.5481	0.6515
CatBoostClassifier	Default	0.7891	0.6978	0.559	0.6699
Grid_CatBoostClassifier	(depth = 8, iterations = 100, learning_rate = 0.04)	0.7981	0.6982	0.5608	0.6648
Rand_CatBoostClassifier	(depth = 8, iterations = 100, learning_rate = 0.04)	0.7981	0.6982	0.5608	0.6648
Tpot_CatBoostClassifier	(depth = 6, iterations = 100, learning_rate = 0.06)	0.8	0.7057	0.5729	0.6764
Grid_LGBMClassifier	(boosting_type = gbdt, colsample_bytree = 0.7, learning_rate = 0.05, max_depth = 8, metric = auc, min_data_in_leaf = 10, min_split_gain = 0.01, num_leaves = 200, objective = binary, random_state = 501, subsample = 0.5)	0.7839	0.6925	0.5503	0.6644

(*continued*)

Table 4. (*continued*)

Name	parameters	Accuracy	Roc_auc	F1score	Gmean
LGBMClassifier	Default	0.7858	0.6983	0.5595	0.6729
Tpot_LGBMClassifier	(boosting_type = gbdt, colsample_bytree = 0.7, learning_rate = 0.05, max_depth = 7, metric = auc, min_data_in_leaf = 10, min_split_gain = 0.01, num_leaves = 200, objective = binary, random_state = 700, subsample = 0.5)	0.7891	0.6995	0.5616	0.6728
Rand_LGBMClassifier	(boosting_type = gbdt, colsample_bytree = 0.7, learning_rate = 0.05, max_depth = 8, metric = auc, min_data_in_leaf = 10, min_split_gain = 0.01, num_leaves = 300, objective = binary, random_state = 700, subsample = 0.5)	0.7882	0.7	0.5622	0.6741
Hyperopt_LGBMClassifier	(boosting_type = gbdt, colsample_bytree = 0.5, learning_rate = 0.05, max_depth = 6, metric = f1, min_data_in_leaf = 10, min_split_gain = 0.01, num_leaves = 90, objective = binary, random_state = 501, subsample = 0.7)	0.7991	0.7057	0.5726	0.6769

Random Forest: Accuracy, ROC-AUC, F1-Score, and G-mean all get better with each hyperparameter setting for random forest models.

XGBoost: For each hyperparameter setting, accuracy, ROC-AUC, and F1-score improve, with the exception of grid search. On the other hand after TPE and genetic optimization techniques, G-mean has increased.

CatBoost: Accuracy increased with every hyperparameter setting, where ROC-AUC and F1-score were enhanced after grid search, random search, and the genetic optimization algorithm. G-mean increased only after the genetic optimization technique was used. The TPE optimization algorithm did not improve any classifier's performance.

LightGBM: After adjusting each hyperparameter with the exception of grid search, accuracy, ROC-AUC, and F1 score all improve. However, G-mean increased after the random search and TPE. Grid search does not perform well for those classifiers.

Here, we show the ROC-AUC in Fig. 3 and F1-Score in Fig. 4 as the performance of classifiers along with hyperparameter tuning.

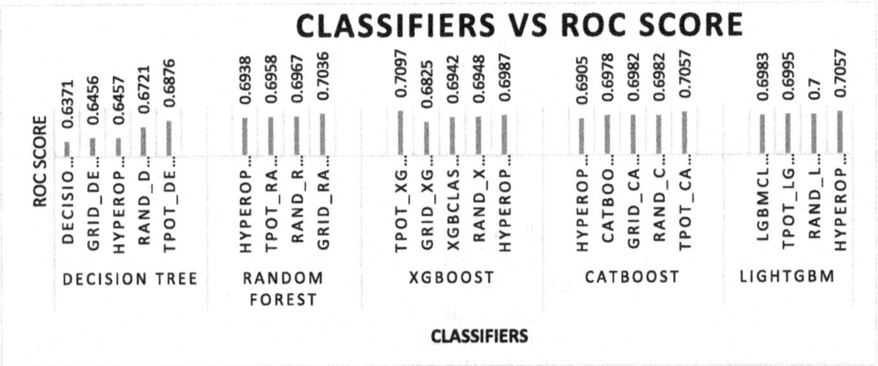

Fig. 3. ROC-AUC expansion for each classifier prior to and following the hyperparameter change.

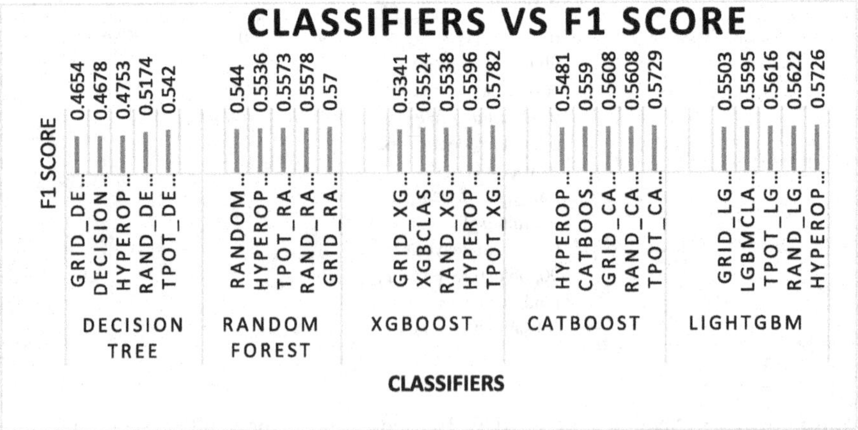

Fig. 4. F1-score of each classifier that employs hyperparameter optimization techniques.

6 Conclusion and Future Scope

As per result of first experiment, The performance of logistic regression and SGD is effectively improved after grid search, random search, TPE, and the genetic algorithm. As a conclusion, KNN, SGD, and LR show good performance for all hyperparameter methods. In second experiment, following the adjustment of all the hyperparameters, the random forest performs well enough when compared to other classifiers. On the other hand, the performance of all classifiers is improved by the genetic optimization approach, which works relatively well. However, the performance of classifiers is affected by a variety of factors, with hyperparameters playing a significant role. Choices of hyperparameter for classification of imbalanced dataset is not an easy tusk, it depends on nature of problem, types of dataset, data complexity and type of classifier etc. Hence application of proper hyperparameter tuning for classification on specific imbalanced data could be major future tusk.

It is challenging to choose hyperparameters for the classification of an unbalanced dataset since they depend on the nature of the problem, the type of dataset, the complexity of the data, the type of model, and other factors. Therefore, applying the right hyperparameter tuning for classification to particular imbalanced data could be a big future tusk. Classification of overlapping imbalanced dataset using hyperparameter optimization could be a another challenging future scope. Hyperparameter based classification of an overlapping, unbalanced dataset may prove to be another challenging task in the future.

One of the most popular methods that researchers take into consideration when dealing with the categorization of unbalanced data is hyperparameter tweaking of the predictive model in addition to conventional resampling. Hyperparameter optimization of resampling techniques has not been widely used by researchers.

Hence, effective hyperparameter tuning of resampling techniques to enhance the classification result is perhaps a future aspect. Another approach to using a pipeline is one in which we combine the predictive model's hyperparameter setting and the resampling technique to achieve an effective result.

References

1. Japkowicz, N., Stephen, S.: The class imbalance problem: a systematic study. Intell Data Anal **6**, 429–449 (2002)
2. Li, P., Yin, L., Zhao, B., Sun, Y.: Virtual screening of drug proteins based on imbalance data mining. Math. Probl. Eng. 10 (2021). Article ID 5585990. https://doi.org/10.1155/2021/5585990
3. Zhang, J., Chen, L., Abid, F.: Prediction of breast cancer from imbalance respect using cluster-based undersampling method. J. Healthc. Eng. (2019)
4. Makki, S., Assaghir, Z., Taher, Y., Haque, R., Hacid, M.-S., Zeineddine, H.: An experimental study with imbalanced classification approaches for credit card fraud detection. IEEE Access (2019). https://doi.org/10.1109/ACCESS.2019.2927266. Advanced Software and Data Engineering for Secure Societies

5. Effendy, V., Adiwijaya, K., Baizal, A.: Handling imbalanced data in customer churn prediction using combined sampling and weighted random forest. In: 2nd International Conference on Information and Communication Technology (ICoICT) (2014). https://doi.org/10.1109/ICoICT.2014.6914086

6. Paing, M.P., Choomchuay, S.: Improved random forest (RF) classifier for imbalanced classi-fication of lung nodules. In: International Conference on Engineering, Applied Sciences, and Technology (ICEAST). IEEE (2018). https://doi.org/10.1109/ICEAST.2018.8434402

7. Roy, K., et al.: An enhanced machine learning framework for type 2 diabetes classification using imbalanced data with missing values. Complexity **2021**, 21. https://doi.org/10.1155/2021/9953314. Article ID 9953314

8. Zhang, X., Zhuang, Y., Wang, W., Pedrycz, W.: Transfer boosting with synthetic instances for class imbalanced object recognition. IEEE Trans. Cybern. **48**(1), 357–370 (2018)

9. Lin, W., Wu, Z., Lin, L., Wen, A., Li, J.: An ensemble random forest algorithm for insurance big data analysis. IEEE Access **5**, 16568–16575 (2017)

10. Rekha, G., Krishna Reddy, V., Tyagi, A.K.: An Earth mover's distance-based undersampling approach for handling class-imbalanced data. Int. J. Intell. Inf. Database Syst. **13**(2–4), 376–392 (2020). https://doi.org/10.1504/IJIIDS.2020.109463

11. Wong, G.Y., Leung, F.H.F., Ling, S.H.: A novel evolutionary preprocessing method based on oversampling and under-sampling for imbalanced datasets. In: IECON 2013—39th Annual Conference of the IEEE Industrial Electronics Society, pp. 2354–2359. IEEE, Vienna (2014). https://doi.org/10.1109/IECON.2013.6699499

12. Kaur, H., Pannu, H.S., Malhi, A.K.: A systematic review on imbalanced data challenges in machine learning: applications and solutions. ACM Comput. Surv. **52**(4), 36 (2019). https://doi.org/10.1145/3343440. Article 79

13. Kong, J., Kowalczyk, W., Nguyen, D.A., Bäck, T., Menzel, S.: Hyperparameter optimi-sation for improving classification under class imbalance. In: IEEE Symposium Series on Computational Intelligence (SSCI) (2019). https://doi.org/10.1109/SSCI44817.2019.9002679

14. Tharwat, A., Gabel, T.: Parameters optimization of support vector machines for imbalanced data using social ski driver algorithm. Neural Comput. Appl. **32**, 6925–6938 (2020). https://doi.org/10.1007/s00521-019-04159-z

15. Akın, P.: A new hybrid approach based on genetic algorithm and support vector machine methods for hyperparameter optimization in synthetic minority over-sampling technique (SMOTE). AIMS Math. **8**(4), 9400–9415 (2023). https://doi.org/10.3934/math.2023473

16. Guido, R., Groccia, M.C., Conforti, D.: Hyper-parameter optimization in support vector machine on unbalanced datasets using genetic algorithms. In: Amorosi, L., Dell'Olmo, P., Lari, I. (eds.) Optimization in Artificial Intelligence and Data Sciences. AIROSS, vol. 8, pp. 37–47. Springer, Cham (2022). https://doi.org/10.1007/978-3-030-95380-5_4

17. Li, F., Zhang, X., Zhang, X., Du, C., Xu, Y., Tian, Y.-C.: Cost-sensitive and hybrid-attribute measure multi-decision tree over imbalanced data sets. Inf. Sci. **422**, 242–256 (2018). https://doi.org/10.1016/j.ins.2017.09.013

18. Shi, Z.: Improving k-nearest neighbors algorithm for imbalanced data classification. IOP Conf. Ser.: Mater. Sci. Eng. **719**, 012072 (2020). https://doi.org/10.1088/1757-899X/719/1/012072

19. Cao, P., Zhao, D., Zaiane, O.: An optimized cost-sensitive SVM for imbalanced data learning. In: Pei, J., Tseng, V.S., Cao, L., Motoda, H., Xu, G. (eds.) PAKDD 2013. LNCS (LNAI), vol. 7819, pp. 280–292. Springer, Heidelberg (2013). https://doi.org/10.1007/978-3-642-37456-2_24

20. Firdous, N., Bhardwaj, S.: Handling of derived imbalanced dataset using XGBoost for iden-tification of pulmonary embolism—A non-cardiac cause of cardiac arrest. Med. Biol. Eng. Comput. **60**, 551–558 (2022). https://doi.org/10.1007/s11517-021-02455-2

21. Wang, Z., Wu, C., Zheng, K., Niu, X., Wang, X.: SMOTETomek-based resampling for personality recognition. IEEE Access **7**, 129678–129689 (2019). https://doi.org/10.1109/ACCESS.2019.2940061

22. Jeni, L.A., Cohn, J.F., De La Torre, F.: Facing imbalanced data–recommendations for the use of performance metrics. In: 2013 Humaine Association Conference on

23. Zhang, X., Li, X., Feng, Y.: A classification performance measure considering the degree of classification difficulty. Neurocomputing **193**, 81–91 (2016)

24. Guido, R., Groccia, M.C., Conforti, D.: A hyper-parameter tuning approach for cost-sensitive support vector machine classifiers. Soft. Comput. (2022). https://doi.org/10.1007/s00500-022-06768-8

25. Sarafianos, N., Xu, X., Kakadiaris, I.A.: Deep imbalanced attribute classification using visual attention aggregation. In: Proceedings of the European Conference on Computer Vision, ECCV, pp. 680–697

26. Barandela, R., Valdovinos, R.M., Sánchez, J.S.: New applications of ensembles of classifiers. Pattern Anal. Appl. **6**(3), 245–256 (2003)

27. H.K. Lee, S.B. Kim, An overlap-sensitive margin classifier for imbalanced and overlapping data, Expert Syst. Appl. (2018)

28. Zhang, F., Petersen, M., Johnson, L., Hall, J., O'Bryant, S.E.: Hyperparameter tuning with high performance computing machine learning for imbalanced Alzheimer's disease data. Appl. Sci. **12**, 6670 (2022). https://doi.org/10.3390/app12136670

29. Burduk, R.: Classification performance metric for imbalance data based on recall and selectivity normalized in class labels. Wroclaw University of Science and Technology (2020)

30. Chawla, N.V., Bowyer, K.W., Hall, L.O., Kegelmeyer, W.P.: SMOTE: synthetic minority over-sampling technique. J. Artif. Intell. Res. **16**, 321–357 (2002)

31. Han, H., Wang, W.-Y., Mao, B.-H.: Borderline-SMOTE: a new over-sampling method in imbalanced data sets learning. In: Huang, D.-S., Zhang, X.-P., Huang, G.-B. (eds.) ICIC 2005. LNCS, vol. 3644, pp. 878–887. Springer, Heidelberg (2005). https://doi.org/10.1007/11538059_91

32. Hu, S., Liang, Y., Ma, L., He, Y.: MSMOTE: improving classification performance when training data is imbalanced. In: 2009 Second International Workshop on Computer Science and Engineering, vol. 2, pp. 13–17 (2009). https://doi.org/10.1109/WCSE.2009.756

33. Mathew, J., Luo, M., Pang, C.K., Chan, H.L.: Kernel-based SMOTE for SVM classification of imbalanced datasets. In: IECON 2015 - 41st Annual Conference of the IEEE Industrial Electronics Society (2016). https://doi.org/10.1109/IECON.2015.7392251

34. He, H., et al.: ADASYN: adaptive synthetic sampling approach for imbalanced learning. In: 2008 IEEE International Joint Conference on Neural Networks (IEEE World Congress on Computational Intelligence), pp. 1322–1328 (2008). https://doi.org/10.1109/IJCNN.2008.4633969

35. Zhu, Y., Jia, C., Li, F., Song, J.: Inspector: a lysine succinylation predictor based on edited nearest-neighbor undersampling and adaptive synthetic oversampling. Anal. Biochem. **593**, 113592 (2020).https://doi.org/10.1016/j.ab.2020.113592

36. Elhassan, A.T., Aljourf, M., Al-Mohanna, F., Shoukri, M.: Classification of imbalance data using tomek link (T-Link) combined with random under-sampling (RUS) as a data reduction method. Glob. J. Technol. Optim. **S1**, 111 (2017). https://doi.org/10.4172/2229-8711.S1:111

37. Rayhan, F., Ahmed, S., Mahbub, A., Jani, R., Shatabda, S., Farid, D.M.: CUSBoost: cluster-based under-sampling with boosting for imbalanced classification. In: 2nd International Conference on Computational Systems and Information Technology for Sustainable Solution, CSITSS 2017, pp. 1–5 (2018). https://doi.org/10.1109/CSITSS.2017.8447534

38. Chen, X., Kang, Q., Zhou, M., Wei, Z.: A novel under-sampling algorithm based on iterative-partitioning filters for imbalanced classification. In: IEEE International Conference on Automation Science and Engineering (CASE), pp. 490–494 (2016). https://doi.org/10.1109/COASE.2016.7743445

39. Rekha, G., Tyagi, A.K., Reddy, V.K.: Performance analysis of under-sampling and over-sampling techniques for solving class imbalance problem. In: International Conference on Sustainable Computing in Science, Technology & Management (SUSCOM). Elsevier (2019)

40. Werner de Vargas, V., Schneider Aranda, J.A., dos Santos Costa, R., et al.: Imbalanced data preprocessing techniques for machine learning: a systematic mapping study. Knowl. Inf. Syst. **65**, 31–57 (2023). https://doi.org/10.1007/s10115-022-01772-8

41. Iranmehr, A., Masnadi-Shirazi, H., Vasconcelos, N.: Cost-sensitive support vector machines **343**, 50–64 (2019). https://doi.org/10.1016/j.neucom.2018.11.099

42. Xuan, P., Sun, C., Zhang, T., Ye, Y., Shen, T., Dong, Y.: Gradient boosting decision tree-based method for predicting interactions between target genes and drugs. Front. Genet. (2019). https://doi.org/10.3389/fgene.2019.00459

43. Zhang, Y., et al.: Research and application of AdaBoost algorithm based on SVM. In: 2019 IEEE 8th Joint International Information Technology and Artificial Intelligence Conference (ITAIC), Chongqing, China, pp. 662–666 (2019). https://doi.org/10.1109/ITAIC.2019.8785556

44. Ogunleye, A., Wang, Q.-G.: XGBoost model for chronic kidney disease diagnosis. IEEE/ACM Trans. Comput. Biol. Bioinform. **17**(6), 2131–2140 (2020). https://doi.org/10.1109/TCBB.2019.2911071

45. Song, J., Lu, X., Wu, X.: An improved AdaBoost algorithm for unbalanced classification data. In: 2009 Sixth International Conference on Fuzzy Systems and Knowledge Discovery, Tianjin, China, pp. 109–113 (2009). https://doi.org/10.1109/FSKD.2009.608

46. Ayyagari, M.R.: Classification of imbalanced datasets using one-class SVM, k-nearest neighbors and CART algorithm. Int. J. Adv. Comput. Sci. Appl. **11**(11) (2020). https://doi.org/10.14569/IJACSA.2020.0111101

47. Vijayakumar, V., Divya, N.S., Sarojini, P., Sonika, K.: Isolation forest and local outlier factor for credit card fraud detection system. Int. J. Eng. Adv. Technol. (IJEAT) **9**(4) (2020). ISSN 2249–8958

48. Probst, P., Boulesteix, A.-L., Bischl, B.: Tunability: importance of hyperparameters of machine learning algorithms. J. Mach. Learn. Res. **20**, 1–32 (2019)

49. Hutter, F., Hoos, H., Leyton-Brown, K.: An efficient approach for assessing hyperparameter importance. In: ICML, Volume 32 of JMLR Workshop and Conference Proceedings, pp. 754–762 (2014)

50. Nocedal, J., Wright, S.: Numerical Optimization. Springer (2006). ISBN 978-0-387-40065-5

51. Yang, L., Shami, A.: On hyperparameter optimization of machine learning algorithms: theory and practice. Neurocomputing **415**, 295–316 (2020). https://doi.org/10.1016/j.neucom.2020.07.061

52. Bischl, B., et al.: Hyperparameter optimization: foundations, algorithms, best practices, and open challenges. WIREs Data Min. Knowl. Discov. e1484 (2023). https://doi.org/10.1002/widm.1484

53. Bergstra, J., Bengio, Y.: Random search for hyper-parameter optimization. J. Mach. Learn. Res. **13**, 281–305 (2012)

54. Karnin, Z., Koren, T., Somekh, O.: Almost optimal exploration in multi-armed bandits. In: Proceedings of the 30th International Conference on Machine Learning (ICML-13), pp. 1238–1246 (2013)

55. Snoek, J., Larochelle, H., Adams, R.P.: Practical Bayesian optimization of machine learning algorithms. In: Proceedings of the 25th International Conference on Neural Information Processing Systems - Volume 2 (NIPS 2012), pp. 2951–2959. Curran Associates Inc., Red Hook (2012)

56. Bengio, Y.: Gradient-based optimization of hyperparameters. Neural Comput. **12**(8), 1889–1900 (2000). https://doi.org/10.1162/089976600300015187

57. Li, L., Jamieson, K., DeSalvo, G., Rostamizadeh, A., Talwalkar, A.: Hyperband: a novel bandit-based approach to hyperparameter optimization. J. Mach. Learn. Res. **18**(1), 6765–6816 (2017)

58. Aszemi, N.M., Dominic, P.D.D.: Hyperparameter optimization in convolutional neural network using genetic algorithms. Int. J. Adv. Comput. Sci. Appl. **10**(6) (2019)

59. Shekhar, S., Bansode, A., Salim, A.: A comparative study of hyper-parameter optimization tools. In: 2021 IEEE Asia-Pacific Conference on Computer Science and Data Engineering (CSDE), Brisbane, Australia, pp. 1–6 (2021). https://doi.org/10.1109/CSDE53843.2021.9718485

60. https://towardsdatascience.com/hyperparameter-tuning-and-sampling-strategy-1014e05f6c14. Accessed 15 June 2021

61. Sun, M., Dou, H., Li, B., Yan, J., Ouyang, W., Cui, L.: AutoSampling: search for effective data sampling schedules. In: Proceedings of the 38th International Conference on Machine Learning. PMLR **139** (2021)

62. Li, J., Fong, S., Mohammed, S., Fiaidhi, J.: Improving the Classification Performance of Biological Imbalanced Datasets by Swarm Optimization Algorithms. Springer, New York (2015). https://doi.org/10.1007/s11227-015-1541-6

63. Sağlam, F., Sözen, M., Cengiz, M.A.: Optimization based undersampling for imbalanced classes. Adiyaman J. Sci. **11**(2), 385–409 (2021)

64. Moniz, N., Monteiro, H.: No Free Lunch in imbalanced learning. Knowl.-Based Syst. **227**, 107222 (2021). https://doi.org/10.1016/j.knosys.2021.107222

65. Nguyen, D.A., et al.: Improved automated CASH optimization with tree Parzen estimators for class imbalance problems. In: 2021 IEEE 8th International Conference on Data Science and Advanced Analytics (DSAA), Porto, Portugal, pp. 1–9 (2021). https://doi.org/10.1109/DSAA53316.2021.9564147

66. Hancock, J., Khoshgoftaar, T.M.: Impact of hyperparameter tuning in classifying highly imbalanced big data. In: 2021 IEEE 22nd International Conference on Information Reuse and Integration for Data Science (IRI), Las Vegas, NV, USA, pp. 348–354 (2021). https://doi.org/10.1109/IRI51335.2021.00054

67. Sharma, S.R., Singh, B., Kaur, M.: Classification of Parkinson disease using binary Rao optimization algorithms. Expert Syst. **38** (2021). https://doi.org/10.1111/exsy.12674

68. Panda, D.K., Das, S., Townley, S.: Hyperparameter optimized classification pipeline for handling unbalanced urban and rural energy consumption patterns. Expert Syst. Appl. **214**, 119127 (2023). https://doi.org/10.1016/j.eswa.2022.119127. ISSN 0957-4174

69. Bertsimas, D., Wang, Y.: Imbalanced classification via robust optimization

70. Rosales-Pérez, A., García, S., Herrera, F.: Handling imbalanced classification problems with support vector machines via evolutionary bilevel optimization. IEEE Trans. Cybern. https://doi.org/10.1109/TCYB.2022.3163974

71. Muntasir Nishat, M., et al.: A comprehensive investigation of the performances of different machine learning classifiers with SMOTE-ENN oversampling technique and hyperparameter optimization for imbalanced heart failure dataset. Sci. Program. **2022**, 17 (2022). https://doi.org/10.1155/2022/3649406. Article ID 3649406

72. Vuttipittayamongkol, P., Elyan, E., Petrovski, A.: On the class overlap problem in imbalanced data classification. Knowl. Base Syst. (2020). https://doi.org/10.1016/j.knosys.2020.106631

73. Kong, J., Kowalczyk, W., Nguyen, D.A., Bäck, T., Menzel, S.: Hyperparameter optimisation for improving classification under class imbalance. In: 2019 IEEE Symposium Series on Computational Intelligence (SSCI), Xiamen, China, pp. 3072-3078 (2019). https://doi.org/10.1109/SSCI44817.2019.9002679

74. Ke, G., et al.: LightGBM: a highly efficient gradient boosting decision tree. In: Proceedings of the 31st International Conference on Neural Information Processing Systems (NIPS 2017), pp. 3149–3157. Curran Associates Inc., Red Hook (2017)

75. Prokhorenkova, L., Gusev, G., Vorobev, A., Dorogush, A.V., Gulin, A.: CatBoost: unbiased boosting with categorical features. In: Proceedings of the 32nd International Conference on Neural Information Processing Systems (NIPS 2018), pp. 6639–6649. Curran Associates Inc., Red Hook (2018)

Smart Parking System Using Arduino and IR Sensor

R. Chawngsangpuii(✉) 🆔 and Angelina Lalruatfeli 🆔

Department of Information Technology, Mizoram University, Aizawl, India
mzut126@mzu.edu.in

Abstract. Internet of Things is used for the smart parking system with Arduino along with IR sensor in this paper. It shows an effective way of using the parking system more efficient and flexible comparing to other existing parking system. If this parking system comes into existence it will make the end users more flexible and comfortable when compared with the existing parking system. It makes the end users more flexible since customers want their way of life to be more secured but want them to be easier. So, this paper would help in providing the best parking system experience like never before. Even in the real-world scenario, it would give the end-users a lot of advantages like reduced fuel consumption and manpower which would result in making many companies to opt this system for their parking slots.

Keywords: Parking System · Arduino · IR Sensors · Internet of Things · vehicles

1 Introduction

There exist a greater need for effective technology for solving many of our day-to-day problems that are evident on the surface when it comes to smart cities. For drivers, one of the most exasperating issues is finding a parking area in crowded areas. More often than not, drivers usually waste precious time and fuel in finding a parking space [1]. The paper intents at creating and designing a smart parking system that efficiently addresses the issues. The existing system does not have the proper security system that the customers need to park their cars [2]. In this paper, users will be notified when they are out of the parking slot about the maximum cars that could be able to park and if the parking slot is full they may wait till the car comes out or else they can move to any other parking slots next to them. In this way, the proposed system makes the user to move from outside of the parking slot which is better than coming inside and searching and going out. This parking system will provide the best security and it will give the best parking experience and a flexible move to the customers park their cars or vehicles in this intelligent parking system slots.

This Smart parking system with Arduino could make the end users more flexible with lesser time and lesser fuel consumption system. It is important and a very useful system to be implemented on real-world scenario which makes life easier and the less complex when compared to the ordinary parking system [3].

P. Das et al. (Eds.): AMRIT 2023, CCIS 1954, pp. 229–235, 2024.
https://doi.org/10.1007/978-3-031-47221-3_20

The main objective of the paper is to provide customers or the users with the best quality, secure, easier, flexible product that is the intelligent parking system. And the companies that implement the parking system do not need to have the security personnel for the parking slots since they need only the devices for controlling which could be done just by sitting from a specific place.

2 Related Works

The architectural design of Arduino-based vehicle parking systems are explained in [4] in which authorization card is distributed for each user that contains the vehicle registration number and other details. Parking gate will be opened and user will be allowed for parking the vehicle in parking space if user is authorized and space is available, otherwise user will not be allowed even though the user is authorized. In this way, parking issue is solved in city areas which also provides security to the vehicle since any unauthorized user will not be allowed to use the parking space. Multi-floor parking also helps in parking vehicles as floors having free space is displayed.

Use of a rotary parking system which utilizes six parking shelves arranged vertically and able to rotate is introduced by authors in [5]. They found a way to utilize narrow space by using rotation-based automated parking system. The use of controller Arduino Uno module that manages all the devices is implemented.

A parking system that uses Arduino components and android mobile application was introduced by the authors in [6] which assists drivers in finding empty parking space based on the remaining empty slots in the parking area. The system can help economy and environment by decreasing fuel consumption, pollution and time in finding parking space.

A parking system that has ability to allow the access and prevent authorized and unauthorized vehicles respectively in parking area is developed by the authors in [7].

New architecture for improved parking of vehicles is proposed by the authors in [8] in which availability parking status of registered vehicles are examined by the sensor-based parking devices, with the intention distributing slots in the parking area without making the system too inflexible for vehicle parking.

3 Problem Statement

This system uses IOT that is the Internet of Things which means providing the best things on this world to be digitalized and which makes the customers to be more happy and efficient than any other products. We live on the world of digitalized things on our day-to- day life and this parking system would also provide the best of digitalization and flexibility to the customers and consumers live on this technology. This parking system would also decrease the amount of fuels that is used and this would control the air pollution all over the country especially in India. The system could result in providing users using this parking system with best parking system and with best security system compared to the existing systems. The intelligent parking system would be one of the solutions for users to park their vehicles safe with automated security systems [9].

4   Design Implementation

The system is implemented with Arduino UNO [10] as depicted in Fig. 1 in which the digital pins are connected to the two IR sensors. A signal from the IR entry sensor is received by Arduino, which controls the servo motor [11] whenever a vehicle is sensed at the entry gate. Based on the car detected at the entry, the motor functions as a check gate, able to allow and disallow vehicles. At the initial point, the number of parking spaces available has been defined. Both the IR sensors are utilized to detect vehicles entering and exiting the parking area. When one IR sensor outside the gate detects a vehicle entering the parking area, the number of parking space available is decreased by one. When one IR sensor [12] inside the gate detects a vehicle leaving the parking area, the number of parking space available is increased by one. The check gate in the parking area incessantly opens and shuts the gate based on the vehicle detection.

The functional requirement are the general requirements needed for the functioning of the system under the execution of the products and the software that has been used to make the system more accurate and makes the hardware components to work properly and easier with requirements.

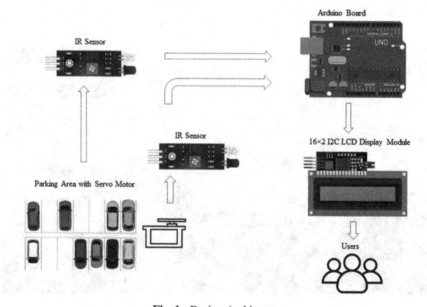

Fig. 1. Design Architecture

The design of any of the Arduino based architecture would generally depend up on the program given through the java coding through the Arduino IDE (Integrated Development Environment). The architectural connections of the pins are generally based on the connectivity of the pins declared on the software. Whenever the compilation is done, it checks whether the board is properly connected or not and also checks with the drivers which could result in some errors if it is not been found on the system, through which the compilation take place and after which compilation is successfully. The next process to be followed after the compilation of the program is the uploading of the program into the Arduino board which gives the specific pins with some data or power as declared in the compiled program. If the hardware components declared on the program and the pins have been connected properly then the program being uploaded will be updated on the board and the hardware works properly.

5 Results and Discussion

The proposed Smart Parking System is found to be capable of detecting the entry or exit of any car, can relay the status of available parking space in addition to saving the data of the IR sensors. Moreover, the improvement of showing actual slot of a parking space to the user is found in our proposed system. Based on the findings, the proposed parking system proved to be an efficient system to develop, and can reduce the various problem of parking system, predominantly the time wasted in searching available spaces for parking. The results showing the parking system is illustrated in Fig. 2, the parking structure of the system that shows the available parking spaces in Fig. 3, the smart parking system when one slot is filled up in Fig. 4, and parking system when parking slots are full in Fig. 5 respectively.

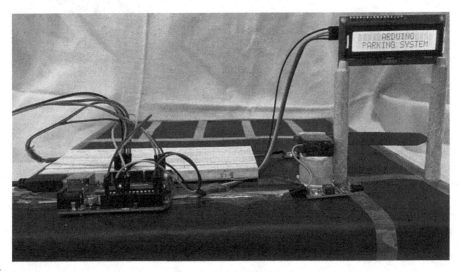

Fig. 2. Parking System

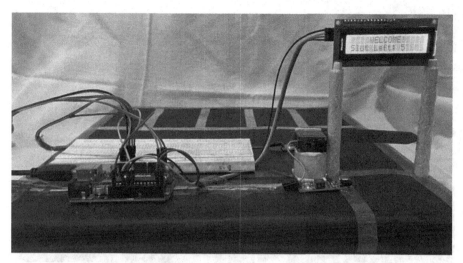

Fig. 3. Parking system showing available Parking slots

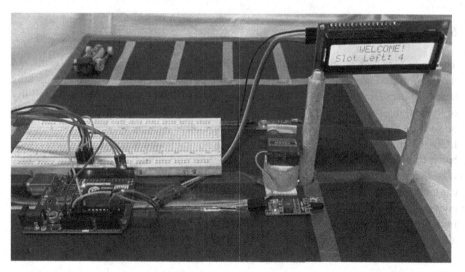

Fig. 4. Parking system with One Slot Filled

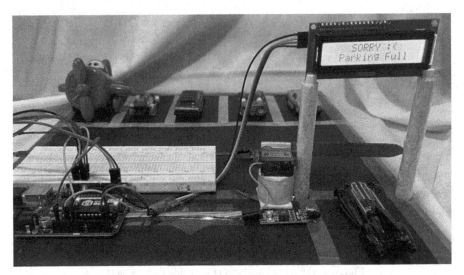

Fig. 5. Parking system when Parking Space is Full

6 Conclusion

The main concepts employed in the proposed parking system are Arduino board and the two infrared sensors based on IoT which offers an efficient way for finding parking spaces. It uses IR sensors for detecting the presence of a car for determining availability of the parking space. Parking spaces are constantly monitored and the updated data can be seen on the LCD screen. Exact location of available parking slot can be seen on the LCD screen. The prototype in this paper was designed for an individual parking space, but it can be expanded for numerous parking spaces. Additionally, a system interface was also developed to record the availability of parking space and the time at which car enters or leaves a parking space. Consequently, the proposed system provides more convenience to drivers or users which allows tackles the problem of wastage in time and fuel consumption. In expansion of this work, a mobile application can be created for allowing users to search, detect and reserve a parking space online.

References

1. Zhou, F., Li, Q.: Parking guidance system based on ZigBee and geomagnetic sensor technology. In: Proceedings of the 13th IEEE International Symposium on Distributed Computing and Applications to Business, Engineering and Science (DCABES), pp. 268–271 (2014)
2. Rahman, S., Bhoumik, P.: IoT based smart parking system. Int. J. Adv. Comput. Electron. Eng. 4(1), 11–16 (2019)
3. Fahim, A., Hasan, M., Chowdhury, M.A.: Smart parking systems: comprehensive review based on various aspects. Heliyon 7(5) (2021). https://doi.org/10.1016/j.heliyon.2021.e07050
4. Chaudhary, H., Bansal, P., Valarmathi, B.: Advanced CAR parking system using Arduino. In: Proceedings of the 4th International Conference on Advanced Computing and Communication Systems (ICACCS), pp. 1–5 (2017). https://doi.org/10.1109/ICACCS.2017.8014701

5. Sodiq, M., Hasbullah, H.: Prototype of Arduino based parking rotation system. In: Proceedings of International Symposium on Materials and Electrical Engineering (ISMEE) (2017). IOP Conf. Ser.: Mater. Sci. Eng. **384**, 012013 (2018). https://doi.org/10.1088/1757-899X/384/1/012013

6. Mahir, K.: Arduino smart parking manage system based on ultrasonic internet of things (IoT) technologies. Int. J. Eng. Technol. **7**(3), 494–501 (2018). Special Issue 20

7. Sabbea, M.O.B., Irfan, M., Ltamimi, S.K.A., Saeed, S.M., Almawgani, A.H.M., Alghamdi, H.: Design and development of a smart parking system. J. Autom. Control Eng. **6**(2), 66–69 (2018). https://doi.org/10.18178/joace.6.2.66-69

8. Kannan, M., Mary, L.W., Priya, C., Manikandan, R.: Towards smart city through virtualized and computerized car parking system using Arduino in the internet of things. In: Proceedings of International Conference on Computer Science, Engineering and Applications (ICCSEA), pp. 1–6 (2020). https://doi.org/10.1109/ICCSEA49143.2020.9132876

9. Ibrahim, H.: Car parking problem in urban areas, causes and solutions. In: Proceedings of the 1st International Conference on Towards a Better Quality of Life (2017). https://doi.org/10.2139/ssrn.3163473

10. Arduino Homepage. https://www.arduino.cc/. Accessed 01 Jan 2023

11. Electrical4U Webpage. https://www.electrical4u.com/what-is-servo-motor/. Accessed 02 Jan 2023

12. STEMpedia Webpage. https://thestempedia.com/tutorials/what-is-an-ir-sensor/. Accessed 02 Jan 2023

QuMaDe: Quick Foreground Mask and Monocular Depth Data Generation

Sridevi Bonthu[1]([✉]) [iD], Abhinav Dayal[1] [iD], Arla Lakshmana Rao[2],
and Sumit Gupta[1] [iD]

[1] Department of CSE, Vishnu Institute of Technology, Bhimavaram, India
sridevi.db@gmail.com, abhinav.dayal@vishnu.edu.in
[2] Service Engineer, Providence India, Hyderabad, India

Abstract. Segmentation of the desired object along with depth estimation is useful in various applications like robotics and autonomous navigation. Any deep learning workflow to estimate monocular depth and segment the desired foreground object in a scene requires significant training data. The data generation process usually involves expensive hardware like RGB-D sensors, laser scanners, or significant manual involvement. Moreover, for every specific foreground object, the data collection process needs to be repeated. This paper presents a novel way to utilize only a small number of readily available png images with transparency for the foreground object and representative background images from the internet and combine them to generate a large dataset for deep learning, utilizing recent monocular depth estimation techniques. To illustrate the effectiveness of the data generation approach, this paper presents a baseline model for depth and foreground mask estimation for detecting cattle on roads using the generated data from the proposed approach. The baseline model exhibits strong generalization to real scenarios. The generated dataset is available for public use.

Keywords: Encoder-Decoder architecture · Dense depth prediction · Mask prediction · custom dataset · autonomous navigation

1 Introduction

Depth estimation and segmentation of desired objects in the scene are often used together in many vision tasks [1], like autonomous navigation of agents, augmented reality, self-driving cars, and other robotics applications. In all these applications, the precise identification of desired objects in the scene and their depth estimate from the camera are crucial for safe and effective navigation. Modern RGB-D sensors like OAK-D[1] are capable of concurrently executing advanced deep neural networks while providing depth and color information from two stereo cameras and a single camera (4K) placed in the centre respectively. Deep learning-based techniques using convolutional neural networks have effective solutions in both depth estimation and semantic segmentation. In general, the

---

[1] OpenCV AI Kit: https://opencv.org/introducing-oak-spatial-ai-powered-by-opencv/.

P. Das et al. (Eds.): AMRIT 2023, CCIS 1954, pp. 236–246, 2024.
https://doi.org/10.1007/978-3-031-47221-3_21

availability of large training datasets is required for a deep learning network to produce high accuracy results. The collection of such data is expensive and time-consuming. Specific applications requiring several foreground objects against a variety of backgrounds become even more challenging in terms of simulating those scenarios. Synthetic datasets using virtual reality have been proposed to that end [2].

Recent research indicates effective use of readily available images on the internet to curate training data [3]. This paper introduces QuMaDe, a way to curate a custom dataset containing hundreds of thousands to millions of images by multiplexing desired foreground objects over representative background scenes, while also generating corresponding depth and foreground mask images. This reduces both the cost and the time required. The authors also experiment by creating baseline models for several application contexts and show that the generated data successfully generalizes to detect relevant objects in real-world scenes. Multiplexing, combined with random cropping, scaling, and translation, makes the data generation fast and effective. With only 100 pairs of background and foreground images, the authors generate 0.4 million triplets of foreground over background, monocular depth, and foreground mask images by effectively leveraging existing SOTA models for depth estimation.

The key contributions of this work are:

1. A novel effort to mix and match foreground and background images, reducing the need for complex scene generation for data curation.
2. Curate a large dataset to effectively train models to detect depth and mask for specific foreground objects over any target background, from a limited input of readily available internet images.
3. Combine image, depth map, and foreground mask in a single dataset using current SOTA models for depth estimation.
4. To release the curated dataset and the trained models, making them publicly available. Researchers can use this single dataset to do segmentation, train models to predict depth, or predict both depth and mask.

Fig. 1. Sample Record (Background image, foreground on background, mask, depth)

The Fig. 1 represents an example from the generated dataset to help detect cattle on roads, a common occurrence on Indian roads that leads to several accidents each year involving loss of life and property. The generated dataset, Monocular Depth Estimation and Segmentation (MODES)[2] and the trained model are publicly available.

2 Related Work

A depth image is a channel of an RGB image in which each pixel corresponds to the separation between the image plane and the related object. Monocular depth gives information related to the depth and distance, and Monocular Depth Estimation (MDE) is the task of approximating the scene depth using a one image [4]. Image segmentation is the method of partitioning an image into various segments that can be used for locating objects and boundaries [5]. A RGBD image is considered as a amalgamation of an RGB image and its equivalent depth image [6].

Depth information is vital to many problems, and it is involved in the tasks like robotics, mapping, localization, obstacle avoidance, autonomous navigation, and computer vision [7]. The datasets of RGBD images are usually collected using depth sensors, monocular cameras, and LiDAR scanners. All the hardware required for the above are expensive and the data curation is also time consuming job. The well known datasets for monocular 3D object detection are Context-Aware MixEd ReAlity (CAMERA), Objectron, Kitty3D, Cityscape3D, Synthia, etc., and these datasets have limitations like indoor-only images, a small number of training examples, and sparse sampling. Some of the most frequently used RGBD datasets are the Kitti [8], the Synthia [2], the Make3D [9], and the NYU [10].

The dataset Kiiti [8] was gathered with the help of a vehicle outfitted with a sparse Velodyne LiDAR scanner (VLP-64) and RGB cameras, and it portrays street scenes in and around the place Karlsruhe, Germany. The primary application of this dataset involves perception tasks in the context of autonomous driving. Synthia [2] is a street scene dataset with depth maps of synthetic data, needing domain adaptation to apply to real world backgrounds. Cityscapes [11] provides a dataset of street scenes, although with more diversity than the images of Kitti. Sintel [12] is one more artificial dataset that mostly includes outdoor scenes. The dataset Megadepth [3] is an extensive set of outdoor images collected from the internet with depth maps reconstructed using structure from the motion techniques. But this dataset lacks ground truth depth and scale. The RedWeb [13] dataset provides depth maps produced from stereo images, which are freely available on large-scale data platforms such as Flickr. The datasets MegaDepth and RedWeb can be effortlessly computed with the existing MVS methods. The dataset Make3D [9] provides the RGB and depth information for outdoor scenes. The NYUv2 [10] dataset is widely used for MDE in the indoor environments. The data was collected using a Kinect RGBD camera, which provides sparse and noisy depth images.

Most of the existing datasets consist of indoor images or outdoor images of city streets. For every specific application, like detecting animals roaming on roads for self-driving or assisted-driving cars, or people inside a room for autonomous room cleaners, etc., researchers need to curate specific datasets to train relevant deep learning models. This paper proposes a technique to come up with a custom dataset by using existing

---

[2] Curated Dataset for Cattle on Road: https://www.kaggle.com/datasets/bsridevi/modes-dataset-of-stray-animals.

accurate depth predictor models, like High Quality MDE via Transfer Learning (nyu.h5) [14], making the task of curating a dataset extremely simple and cost-effective.

3 Method

The curated dataset must have following objectives:

1. It should always include the foreground object.
2. It should drive deep learning models that generalize well.
3. It should provide accurate dense depth maps in line with SOTA models.
4. It should provide accurate foreground mask.
5. It should be able to generate the data online during training phase dynamically.

3.1 Data Acquisition

The first step to curating data is to determine the target application scenario and, thus, the foreground object(s) and representative background context. At the same time, the dataset must have sufficient variability to include the majority of the types and views that the trained deep network model may see when deployed.

Fig. 2. Scene and Foreground object images

We propose to download or take an RGB image of n foreground object(s) and m background images (we used $n = m = 100$), balancing the types and views. For example, for the MODES dataset, we chose several cow, bull, and calf types, individual or in group, sitting, standing, or walking, and from various angles. Similarly for backgrounds, we chose backgrounds of streets, storefronts, main roads, highways, markets, railway tracks, landscapes, garbage piles, etc. PNG images with transparency are readily available on the internet for almost any desired foreground object. Such images will easily allow one to generate a foreground mask from non-transparent pixels. If not, tools like GIMP [15] combined with deep learning foreground extractors[3] can help generate the required PNG foreground images. Figure 2 shows a few of the sample scene and foreground images used for the creation of this dataset.

---

[3] https://www.remove.bg/ uses a combination of Image based techniques and DNN to separate foreground from background.

3.2 Multiplexing and Depth Generation

This step is to place each foreground object several times on the background images, generating an fg-bg image, and The foreground mask corresponds to the placement and scale of the foreground. Depth is computed from the fg-bg image via the model proposed by Ibraheem Alhashim et al. in their paper titled "High Quality MDE via Transfer Learning" [14]. The implementation of the paper is available at git[4]. Because this model requires 448×448 size images as input, we resize all background images to this size while maintaining their aspect ratio.

Algorithm 1: Generate Dataset(*[bgimages], [fgimages],* k, b)

---

input : m Background Image paths, $2n$ Foreground Image paths, multiplexing factor k, batch size b *must be multiple of k*

output: Yield $2kmn$ fg-bg, mask and depth images in batches of size b

for offline use it creates 3 folders with fg_bg, mask and depth each having $2kmn$ images;

for $bg \leftarrow 1$ **to** m **do**
  for $fg \leftarrow 1$ **to** $2n$ **do**
    for $i \leftarrow 1$ **to** k **do**
      $croppedbg \leftarrow$ take maximal random crop of 448×448 from bg without affecting the aspect ratio;
      randomly pick a center point (x, y) in range $[0, 447]$;
      randomly pick a scale in range $[0.3, 0.6]$ (ratio of area fg covers bg);
      create $fg - bg$ image by resizing the fg to scale and place it on top of $croppedbg$ centered at x, y calculated;
      calculate binary mask from current placement of fg by thresholding the transparency channel. save $fg - bg$ image and mask add $fg - bg$ image to a batch;

    if b *new fg-bg images generated* **then**
      run depth model on batch and save corresponding depth images.

Fig. 3. Three image sets resulted from the algorithm used. (top) A scene image on which a foreground object is positioned at random location with random scale, (middle) respective mask for the foreground, and (bottom) calculated depth by using a model.

The data generation process is completely online and produces one batch of images for training a deep model. By randomly repeating one foreground object k times at varied locations and scales for each background image and repeating another k times with a horizontally flipped version of the same foreground, one can generate 2kmn fg-bg

---

[4] Source code for Depth Estimation model: https://github.com/ialhashim/DenseDepth/blob/master/DenseDepth.ipynb.

images. In addition, a random crop from the background image instead of a fixed initial crop from the source image can add more variability to the input. For k = 20 and n = m = 100, this becomes 400,000 fg-bg images. Algorithm 1 describes the data generation process. Figure 3 shows a set of sample images generated by using the Algorithm 1.

4 Experimental Analysis

In this section, we provide a baseline for MDE and foreground segmentation on the generated MODES depth images dataset. CNNs are progressive in exploring structural features and spatial image formation. To come up with a baseline, we started training simple CNN, ResNet, and Unet + +. The advanced models for image segmentation are variations of U-Net and convolutional neural networks [16]. Long skip connections are utilized to skip the features from the diminishing path to the escalating path in order to recover the compressed spatial information lost during downsampling [17]. Short skip connections can be used to build deep FCNs.

4.1 Model

By using both the long and short skip connections, we proposed a light-weight model following the U-Net architecture with two decoder networks meant for foreground mask prediction and depth prediction. The architecture of the model is shown in Fig. 4. The total number of parameters in this model is 5,525,568 including both decoders.

The encoder part of the network is comprised of four downsampling units. Every downsampling unit compresses the input scene image with the help of a series of convolutional operations. In our implementation, the source image of size 128×128, changed into $64 \times 64 \rightarrow 32 \times 32 \rightarrow 16 \times 16 \rightarrow 8 \times 8$. This model has a DepthDecoder and a MaskDecoder, and each of them is comprised of four upsampling units. Atrous and transposed convolution operations are used to expand the compressed source image. The encoder outcome 8×8 is expanded into $16 \times 16 \rightarrow 32 \times 32 \rightarrow 64 \times 64 \rightarrow 128 \times 128$. As shown in model architecture Fig. 4, the outcomes of encoder downsampling units were added to the outcomes of decoder upsampling units.

We have trained this model on the entire MODES dataset from scratch with a train-test-split of 70–30%. During training, the network is trained with a batch-size of 64 for 10 epochs using the SGD optimizer [18]. Every epoch took one hour of time on the GPU because of the huge training data. We have used the OneCycleLR scheduler [19] with a maximum learning rate of 0.1. This made the initial learning rate 0.0099. The Deep Convolutional Neural Networks encoder is fed with an image (128×128), and the first decoder outputs a mask image and the second decoder outputs a depth image. To reduce overfitting [20], and achieve generalization, this work employed the augmentation techniques of Random Rotation, Random Grayscale, Color Jitter, random horizontal flips and random channel swaps.

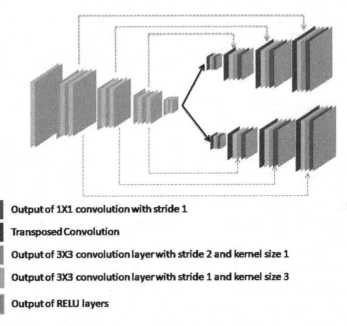

Output of 1X1 convolution with stride 1

Transposed Convolution

Output of 3X3 convolution layer with stride 2 and kernel size 1

Output of 3X3 convolution layer with stride 1 and kernel size 3

Output of RELU layers

Fig. 4. Network Architecture.

4.2 Loss Function

Deciding a universal loss function is not possible for complex objectives like foreground segmentation and depth prediction. Based on the survey [21], we have picked L1-loss and SSIM (Structural Similarity Index) loss [22]. Their work also suggested using a penalty term, which helps the network focus on hard-to-segment boundary regions. The loss is calculated with the help of L1 and SSIM at both decoders and employs regularization for the weight penalty.

For training our network with two decoders, we defined the same loss function L for depth and mask prediction, between y and $\hat{y}$, as the weighted sum of two loss function values.

$$L(y, \hat{y}) = \lambda L_{term1}(y, \hat{y}) + (1 - \lambda)L_{term2}(y, \hat{y})$$

The first loss term, $L_{term1}(y, \hat{y})$, is the point-wise L1-loss defined on the predictions of the Mask Decoder and Depth Decoder units of the network.

$$L_{term1}(y, \hat{y}) = \frac{1}{n} \sum_{x=1}^{n} |y_i - \hat{y_i}|$$

The second loss term $L_{term2}(y, \hat{y})$ uses a commonly used metric for image reconstruction tasks, i.e., SSIM. This metric was used by many recent depth prediction CNNs. The loss term is redefined as shown in equation as SSIM has an upper bound of one.

$$L_{term2}(y, \hat{y}) = \frac{1 - SSIM(y, \hat{y})}{2}$$

Different weight parameters λ were tried and we have ended with a value $\lambda = 0.84$. The final loss function is as follows.

$$L(y, \hat{y}) = 0.84 * L_{term1}(y, \hat{y}) + 0.16 * L_{term2}(y, \hat{y})$$

4.3 Evaluation

To quantitatively measure the efficiency of the proposed model on the MODES dataset, we used seven methods employed in prior work [9]. The error metrics are Absolute Relative Difference, Squared Relative Error, Root Mean Square Error (RMSE), log-RMSE, and Threshold accuracy. The employed metrics are defined as follows.

$$AbsoluteRelativeDifference = \frac{1}{|D|} \sum_{y_p \in D} \frac{|y_p - \widehat{y_p}|}{y_p}$$

$$SquaredRelativeError = \frac{1}{|D|} \sum_{y_p \in D} \frac{\|y_p - \widehat{y_p}\|^2}{y_p}$$

$$RootMeanSquaredError = \sqrt{\frac{1}{|D|} \sum_{y_p \in D} \|y_p - \widehat{y_p}\|^2}$$

$$log - RMSE = \sqrt{\frac{1}{|D|} \sum_{y_p \in D} \|\log(y_{p)} - \log(\widehat{y_p})\|^2}$$

$$Thresholdaccuracy(\delta_t) = \frac{1}{|D|} \left| \left\{ y_p \in D | \max\left(\frac{y_p}{\widehat{y_p}}, \frac{\widehat{y_p}}{y_p} \right) < 1.25^t \right\} \right| \times 100\%$$

where y_p and $\widehat{y_p}$ are pixels in the depth image y and predicted image $\hat{y}$ respectively, D represents set of pixels in the predicted image. δ_t represents the threshold accuracy.

4.4 Results

The model was trained on the entire dataset and obtained significant accuracy with minimal loss. The qualitative performance of the model is depicted through Figs. 5 and 6. The outcome of the model on the validation dataset is shown in Fig. 5. And the unseen data is shown in Fig. 6. The unseen data fed to the model is an actual image, not one that has been curated, as in QuMaDe, where foreground is placed on background. The obtained depths and foreground masks show that the model generalised well. Few exceptions, like the spots on the cow and the two calves not having very good detection, indicate the non-presence of such examples in the training set. However, it should be straightforward to introduce a few more examples to make the application more robust. The Quantitative performance of the model on MODES is presented in the Table 1. They are on par with performance of the models trained on manually annotated data.

Fig. 5. Foreground mask and depth inference on validation data

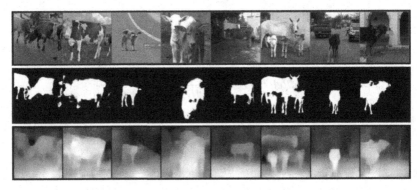

Fig. 6. Foreground mask and depth inference on real life unseen data.

Table 1. Performance metrics of the proposed model on MODES dataset.

Metric	Value
Absolute Relative Difference	0.138
Squared Relative Error	0.499
RMSE	0.059
Log-RMSE	0.818
Threshold$(\delta) < 1.25$	0.790
Threshold$(\delta) < 1.25^2$	0.955
Threshold$(\delta) < 1.25^3$	0.990

5 Conclusion

We presented a novel approach to generate a robust large dataset for depth and foreground mask estimation, where the dataset creation cost is kept low by intelligently multiplexing few foreground objects and high-quality scene images collected from the Internet. We demonstrated the use of generated data to predict depth and mask for cattle on road.

The light-weight baseline model presented, generalizes well to real life unseen data. We also demonstrate successful use of State-of-the-art models like Dense Depth to generate depth images for the curated foreground-background combined images. In addition, we release the 400K multiplexed images, depth, and masks in MODES data for public use.

References

1. Lin, Z., Cohen, S.D., Wang, P., Shen, X., Price, B.L.: U.S. Patent No. 10,019,657. Washington, DC: U.S. Patent and Trademark Office (2018)
2. Ros, G., Sellart, L., Materzynska, J., Vazquez, D., Lopez, A.M.: The synthia dataset: A large collection of synthetic images for semantic segmentation of urban scenes. In: Proceedings of the IEEE Conference on Computer Vision and Pattern Recognition, pp. 3234–3243 (2016)
3. Li, Z., Snavely, N.: Megadepth: learning single-view depth prediction from internet photos. In: Proceedings of the IEEE Conference on Computer Vision and Pattern Recognition, pp. 2041–2050 (2018)
4. Abuolaim, A., Brown, M.S.: Defocus deblurring using dual-pixel data. In: Vedaldi, A., Bischof, H., Brox, T., Frahm, J.-M. (eds.) ECCV 2020. LNCS, vol. 12355, pp. 111–126. Springer, Cham (2020). https://doi.org/10.1007/978-3-030-58607-2_7
5. Amza, C.: A review on neural network-based image segmentation techniques. De Montfort University, Mechanical and Manufacturing Engineering. The Gateway Leicester, LE1 9BH, United Kingdom, vol. 1, p. 23 (2012)
6. Zhang, Y., Funkhouser, T.: Deep depth completion of a single rgb-d image. In: Proceedings of the IEEE Conference on Computer Vision and Pattern Recognition, pp. 175–185 (2018)
7. Marchand, E., Uchiyama, H., Spindler, F.: Pose estimation for augmented reality: a hands-on survey. IEEE Trans. Visual Comput. Graph. **22**(12), 2633–2651 (2015)
8. Geiger, A., Lenz, P., Stiller, C., Urtasun, R.: Vision meets robotics: the kitti dataset. Int. J. Rob. Res. **32**(11), 1231–1237 (2013)
9. Saxena, A., Sun, M., Ng, A.Y.: Make3D: depth perception from a single still image. In: AAAI, vol. 3, pp. 1571–1576 (2008
10. Silberman, N., Hoiem, D., Kohli, P., Fergus, R.: Indoor segmentation and support inference from rgbd images. In: Fitzgibbon, A., Lazebnik, S., Perona, P., Sato, Y., Schmid, C. (eds.) ECCV 2012. LNCS, vol. 7576, pp. 746–760. Springer, Heidelberg (2012). https://doi.org/10.1007/978-3-642-33715-4_54
11. Cordts, M., et al.: The cityscapes dataset for semantic urban scene understanding. In: Proceedings of the IEEE Conference on Computer Vision and Pattern Recognitio, pp. 3213–3223 (2016)
12. Mayer, N., et al.: A large dataset to train convolutional networks for disparity, optical flow, and scene flow estimation. In: Proceedings of the IEEE Conference on Computer Vision and Pattern Recognition, pp. 4040–4048 (2016)
13. Xian, K., et al.: Monocular relative depth perception with web stereo data supervision. In: Proceedings of the IEEE Conference on Computer Vision and Pattern Recognition, pp. 311–320 (2018)
14. Alhashim, I., Wonka, P.: High quality monocular depth estimation via transfer learning. arXiv preprint arXiv:1812.11941 (2018)
15. Howat, I.M., Negrete, A., Smith, B.E.: The Greenland Ice Mapping Project (GIMP) land classification and surface elevation data sets. Cryosphere **8**(4), 1509–1518 (2014)
16. Drozdzal, M., Vorontsov, E., Chartrand, G., Kadoury, S., Pal, C.: The importance of skip connections in biomedical image segmentation. In: Carneiro, G. (ed.) LABELS/DLMIA - 2016. LNCS, vol. 10008, pp. 179–187. Springer, Cham (2016). https://doi.org/10.1007/978-3-319-46976-8_19

17. Zhou, Z., Siddiquee, M.M.R., Tajbakhsh, N., Liang, J.: Unet++: redesigning skip connections to exploit multiscale features in image segmentation. IEEE Trans. Med. Imaging **39**(6), 1856–1867 (2019)

18. Bottou, L.: Large-scale machine learning with stochastic gradient descent. In: Proceedings of COMPSTAT'2010: 19th International Conference on Computational Statistics, Paris, France, 22–27 August 2010, Keynote, Invited and Contributed Papers, pp. 177–186. Physica-Verlag HD (2010)

19. Smith, L.N.: A disciplined approach to neural network hyper-parameters: Part 1--learning rate, batch size, momentum, and weight decay. arXiv preprint arXiv:1803.09820 (2018)

20. Perez, L., Wang, J.: The effectiveness of data augmentation in image classification using deep learning. arXiv preprint arXiv:1712.04621 (2017)

21. Jadon, S.: A survey of loss functions for semantic segmentation. In: 2020 IEEE Conference on Computational Intelligence in Bioinformatics and Computational Biology (CIBCB), pp. 1–7. IEEE (2020)

22. Zhao, H., Gallo, O., Frosio, I., Kautz, J.: Loss functions for neural networks for image processing. arXiv preprint arXiv:1511.08861 (2015)

Fine-Grained Air Quality with Deep Air Learning

Jyoti Srivastava$^{(\boxtimes)}$ [iD], Neha Singh, Rahul Chakravorty, and Anu Raj

Department of ITCA, Madan Mohan Malviya University of Technology, Gorakhpur, India
`sriv.jyoti1996@gmail.com`

Abstract. This research paper **assesses** fine-acquired air quality, which are three various subjects in metropolitan air enlistment. The reactions for these subjects can furnish fundamental data to support with coursing contamination control and, like this, make phenomenal social and specific effects. Latest work manages the three issues independently by various models. One model, Deep Air Learning, the study suggests a fantastic and all-encompassing way to handle the three challenges. The foundation of the DAL philosophy is the integration of semi-controlled learning and highlight choice into multiple tiers of the massive learning affiliation. The suggested approach uses knowledge about unlabelled spatio-transient data to work on the presentation of development and measurement, and it conducts choice and association assessment to uncover the typical big highlights to the arrangement of the air quality. We measure our methods through thorough audits based on reliable information sources located in Beijing, China. Evaluations reveal that DAL outperforms its partner models in terms of resolving issues with expansion, gauge, and component assessment of fine-grained air quality.

Keywords: Air Quality · Metropolitan · Various Models · Deep Air Learning · Forecast

1 Introduction

The development, supposition, and feature examination of precisely measured wind quality are three critical areas that require consideration in the field of urban air registering [1–3]. An effective defense tackles the issue of the city's uneven distribution of scarce air-quality measuring base stations; a precise forecast offers vital information to stop people from being harmed by wind pollution; and a reasonable component examination identifies the critical elements that determine the range of air quality [4, 5]. "When in doubt, the responses for these subjects can isolate incredibly accommodating information to assist with circulating tainting control, and consequently make exceptional social and specific effects [6, 7]. Regardless, there exist a couple of troubles for metropolitan wind figuring as the associated data have few exceptional qualities [8]. Whenever there are a smaller number of wind quality measuring stations in a city it is because of huge cost of made and staying aware of like a station, so it is sometimes costly to conduct checked planning tests when checking fine obtained air quality[9]. Secondly, the identified data

P. Das et al. (Eds.): AMRIT 2023, CCIS 1954, pp. 247–255, 2024.
https://doi.org/10.1007/978-3-031-47221-3_22

of the stations that monitor the air quality are fragmented, and there are various missing signs in the recorded data for particular time periods for specific stations [10]. The air quality screen devices are connected to the legitimization of the missing names. Overall, every station only has one screen device that needs to be observed at distances; as a result, there won't be any output for the station when the apparatus is being observed, calibrated, or experiencing a specific problem[11–13].Thirdly, these city-related air data are varied for the improvement of data throughput. Whatever the case, everything is available to identify the primary cause of the appearing and spread of wind pollution, particularly the pollution of PM 2.5 [14–16]. It is therefore difficult to determine what types of data are the vital components for addition and assumption, as well as the essential parts for environment section to Overcome and regulate air pollution.

1.1 Objective

Objective of this thesis is to a general and effective approach to solve the three problems in one model called the Deep Air Learning (DAL). The main idea of DAL lies in embedding feature selection and semi-supervised learning in different layers of the deep learning network.

The literature review details are found in Sect. 2 of this paper, the approach's details are found. The findings are presented in Sect. 3, followed by a summary of the findings in Sect. 4, and a conclusion in Sect. 5.

2 Related Work

Based on extensive literature survey related to Fine-grained Air Quality with Deep Air Learning has been taken into consideration in this section.

Zhiwen Hu et al. [17] (2019) suggest the development, implementation, and enhancement of their own air quality detection system, which provides a continuous and accurate atmosphere quality guide for the observed area. Four layers of engineering are put up for the framework plan: an energy-efficient detecting layer, a highly reliable transmission layer, a fully highlighted handling layer, and an easily understandable show layer. They have been implementing this system at Peking University (PKU) for a considerable amount of time, and they have collected more than 100,000 information values from 30 devices as part of the execution. Our grounds' landscape is thought to be sufficiently complex to address the common urban area of a huge, bright metropolis because it fully incorporates tall structures, green spaces, and vehicular pathways.

Qi Zhang et al. (2020) [18] suggested a demanding research area would be the spatially precise estimate of wind pollution. The approaches indicated were only preapred for fine-grained wind pollution prediction for the sites (grids) where monitoring systems had provided air pollution data, and they were unable to anticipate places (grids) where monitoring stations were not accessible. One solution to this problem is to use transportable ancillary sensors.

In [19], the authors conducted 3D air quality monitoring using an ELM-based neural network to analyze the data. Although the infrastructure deployment could be very expensive, these methodologies had produced some excellent outcomes for geographically fine-grained wind pollution assessment.

Zhiwen Hu. et al. (2019) [20] use the data acquired by their own framework to represent the estimation error and the induction error in their study. By carefully arranging the force control and area determination procedures, by combining continuous and fine-grained air quality recommendations, we can hopefully lower their joint inaccuracy. The force control problem is approached using a method that relies. By using k-implies bunching for introduction and the hereditary calculation for development, the area determination problem is solved. The two arrangements accomplish good imperfect results, and the blend of them shows a huge prevalence over lessen the normal joint blunder.

Jingchang Huang et al. (2018) [21] A growing number of city dwellers, particularly those who reside in the vast urban networks that are destroyed by air pollution, should consider the significance of air quality to their success. The air quality information is also limited by the additional sense centres, which prompt real requests for considerable standard air quality data gathering. In this study, the authors propose the use of freely supported cars in cities as intrinsic sensors to provide constant and well-sourced air quality data, which essentially increases the plausibility and applicability of the recognition system. When a car's windows are open, the conventional wisdom regarding this work is prompted by the idea that the obsession with air section inside a car is essentially the same as its surrounding environment. Since the obsession design is converged after opening the windows, this paper first develops a skillful estimation to recognise the vehicular wind exchange state, then isolates the intermingling of defilement under that circumstance, and finally denotes the distinguished concurrent worth as a similar air quality level of the including climate. A lavishly displayed city-level air quality aid may be exhibited once progressive wind quality data streams from around the city have been collected and processed in their server ranch using their IoT cloud stage. Evaluations of the computation's display in comparison to the actual data demonstrate the utility of the suggested structure for gathering air quality data under urban situations.

Yuzhe Yang et al. (2019) [22] describes the development of AQNet, a framework for an ethereal ground. They provide the PM 2.5 map via a website-based GUI for constant analysis to ease consumer inquiries. Peking University has acknowledged and sent AQNet nearby, and it is adaptable and energy-compelling to be opened out to larger and more generous districts.

3   Methodology

3.1 Proposed System

We cultivate a general and amazing procedure assembled DAL to tie the addition, assumption, incorporate decision and assessment into a single model of fine-grained air quality. "We suggest a clever method for performing feature assurance in the neural association's data layer, whose headway is manageable and implementation is successful in identifying the key features. The suggested decision and assessment incorporation approach can reveal some of the inner workings of the disclosed important implications models. To accomplish spatio-transient semi-oversaw learning in the neural association, we make use of the quality of the spatio-transitory information as well as the information present in the unlabeled data sample. The DAL model appears to work well in connection with the companion strategies in the composition, according to extensive evaluations on reliable datasets."

3.2 Proposed System Process Flow

1. Input design is the process of converting a user-oriented description of the input into a computer-based system. This design is important to avoid errors in the data input process and show the correct direction to the management for getting correct information from the computerized system.
2. It is achieved by creating user friendly screens for the data entry to handle large volume of data. The goal of designing input is to make data entry easier and to be free from errors. The data entry screen is designed in such a way that all the data manipulates can be performed. It also provides record viewing facilities.
3. When the data is entered it will check for its validity. Data can be entered with the help of screens. Appropriate messages are provided as when needed so that the user will not be in maize of instant (Fig. 1).

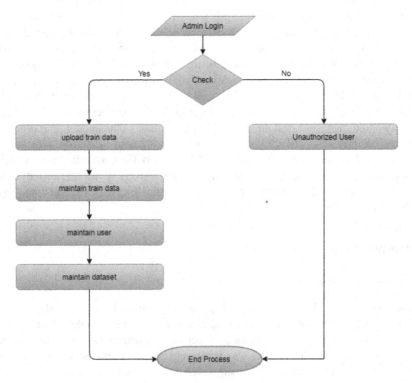

Fig. 1. Admin Flow Diagram

A. **Deep Air Learning**

We proposed a theory known as "Deep Air Learning," in which attribute decision and spatio-short-lived semi-supervised learning are separately integrated in the data layer and the yield layer of the substantial learning network (DAL). For spatial and transient estimation, there is a substantial number of unlabelled data that can be used to pretrain the significant model. [3].

B. **Air quality data**

According to 35 ground-based air quality screen sites, we compile trustworthy centralization of six various types of air pollutants [23]. There are many frameworks that employ information mining to the creation, measurement, and element assessment of air pollution control. It investigates spatio-common inclusion methods for employing air tainting analysis as a development tool. By using a co-planning-based approach, it infers the changing and precise air quality information throughout a city.

C. **Feature Selection**

In various tiers of the substantial learning connection, feature decision and spatio-brief semi-oversaw adjustment were done simultaneously. To address the two topics, we employ a various yield classifier, considering that both the subjects of expansion and assumption are request issues with diverse yields [24]. In this work, present a new substantial learning network such as a novel yield model that uses information identifying with the unlabelled spatio-common information not only to achieve the justification for inclusion but also to erode the introduction of the supposition. [25] By incorporating feature assurance into the planned structure and carrying out connection analysis, the fundamental features that are vital to the assortment of the air quality can also be shown.

D. **k-means Clustering Algorithm**

Signal processing is the initial step in data mining's pack analysis, which uses the well-known vector quantization approach K-implies grouping. As a result, Voronoi cells are created within the data space. Even though the task is computationally difficult (NP-hard), practical heuristic estimations are commonly used and easily reach a local ideal. Both k-infers and Gaussian combination depictions use an iterative refinement method, which is equivalent to the assumption intensification computation for mixes of Gaussian dispersals. Both use cluster centres to present the data, however k-infers clustering will find clusters of similar spatial degree, whereas the assumption helps framework lets clusters have different shapes. [26] The computation employs a well-known AI method for requests that are frequently confused with k-suggests due to the name's k. It is in a free association with the classifier for k-nearest neighbors. On the pack locations acquired by k-expects, one can use the 1-nearest neighbour classifier to integrate new data into the current groups. The Rocchio computation or nearest centroid classifier is used for this.

E. **Image Processing Techniques**

The main purpose of image handling implies automated image planning, i.e., eliminating any turbulence and irregularities from an image utilizing a powerful computer. The disturbance or irregularity could creep into the image as it changes, moves, etc. [27]. The prospective force gains off are generally restricted, discrete sums, we say that an image is automated. A certain number of components, each with a predetermined place and value, must be assembled to create a mechanised image. The terms pixels, pels, picture pieces, and picture parts are used to describe these elements. Pixel is the term used most frequently to define the elements of an automated image.

4   Results

Stn_code	Sampling Data	State	Location	Agency	Type	SO2	NO2	RSPM	SPM	LMS	PM25	Date
150	February-M021990	Andhra Pradesh	Hyderabad	NA	Residential, Rural and other Areas	4.8	17.4	NA	NA	NA	NA	2/1/1990
151	February-M021990	Andhra Pradesh	Hyderabad	NA	Industrial Area	3.1	7	NA	NA	NA	NA	2/1/1990
152	February-M021990	Andhra Pradesh	Hyderabad	NA	Residential, Rural and other Areas	6.2	28.5	NA	NA	NA	NA	2/1/1990
150	March-M031990	Andhra Pradesh	Hyderabad	NA	Residential, Rural and other Areas	6.3	14.7	NA	NA	NA	NA	3/1/1990

Component affirmation and assessment's main objectives are to understand how different data features relate to theories underlying neural associations, identify the components that are pertinent to the various types of air quality, and present research evidence in support of the neutralization and control of air tainting rather than to increase the accuracy of the figure. Most previous feature assurance techniques similarly reduced the gauge precision in varied uses. Our suggested model is flexible, though: when the internet-based structure requires high precision, the part assurance objectives can be dropped; when it's necessary to understand how the framework makes an assumption, contributing the component decision and assessment to the primary estimate technique reveals some internal configuration of the black box. Table 1: Shows the data set (Figs. 2 and 3).

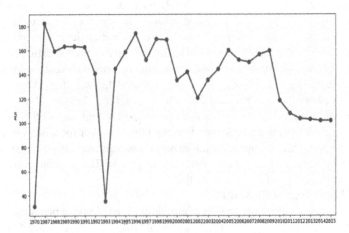

Fig. 2. AQI Variation year wise

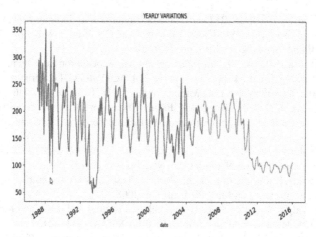

Fig. 3. Yearly Variations

5 Conclusion

The answers to these questions can provide vital information to aid in the control of broadcast defilement and have a profound social and local impact. The majority of present-day projects freely address the three challenges by creating various models. In this research, we support DAL, a comprehensive and practical method that integrates fine-grained air quality monitoring and decision-making into a single model. In order to handle the presentation of expansion and figure, we apply the typical ascribes of the spatio-common data that are collected along with the unlabeled data using spatio-transient semi-supervised learning to the output layers of neural association. Alliance analysis, and with feature assurance, finds the connection between different data aspects and the assumptions underlying brain relationships. In a similar vein, we suggest a brilliant technique for remember decision-making for the neural association's data layer, whose upgrading isn't difficult to address, and execution works well in eliminating the repetitive or superfluous aspects.

References

1. Tibshirani, R.: Regression shrinkage and selection via the lasso. J. Roy. Stat. Soc. Ser. B (Methodol.) **58**(1), 267–288 (1996)
2. Yuan, M., Lin, Y.: Model selection and estimation in regression with grouped variables. J. Roy. Stat. Soc. Ser. B (Stat. Methodol.) **68**, 49–67 (2006)
3. Qi, Z., Wang, T., Song, G., Hu, W., Li, X., Zhang, Z.: Deep air learning: interpolation, prediction, and feature analysis of fine-grained air quality. IEEE Trans. Knowl. Data Eng. **30**(12), 2285–2297 (2018)
4. Dong, M., Yang, D., Kuang, Y., He, D., Erdal, S., Kenski, D.: PM2.5 concentration prediction using hidden semi-markov model-based times series data mining. Expert Syst. Appl. **36**(5), 9046–9055 (2009)
5. Thomas, S., Jacko, R.B.: Model for forecasting expressway pm2.5 concentration – application of regression and neural network models. J. Air Waste Manag. Assoc. **57**(4), 480–488 (2007)

6. Zheng, Y., Liu, F., Hsieh, H.-P.: U-air: when urban air quality inference meets big data. In: Proceedings of the 19th ACM SIGKDD International Conference on Knowledge Discovery and Data Mining, Series KDD 2013, pp. 1436–1444 (2013)

7. Hsieh, H.-P., Lin, S.-D., Zheng, Y.: Inferring air quality for station location recommendation based on urban big data. In: Proceedings of the 21th ACM SIGKDD International Conference on Knowledge Discovery and Data Mining, Series KDD 2015, pp. 437–446 (2015)

8. L. Li, X. Zhang, J. Holt, J. Tian, and R. Piltner, "Spatiotemporal interpolation methods for air pollution exposure," in Symposium on Abstraction, Reformulation, and Approximation, 2011

9. Zhou, X., et al.: Probabilistic dynamic causal model for temporal data. In: 2015 International Joint Conference on Neural Networks (IJCNN), pp. 1–8 (2015)

10. Weston, J., Ratle, F., Collobert, R.: Deep learning via semisupervised embedding. In: the 25th International Conference on Machine Learning (2008)

11. Singh, K.P., Gupta, S., Rai, P.: Identifying pollution sources and predicting urban air quality using ensemble learning methods. Atmos. Environ. **80**, 426–437 (2013)

12. Rosenberg, C., Hebert, M., Schneiderman, H.: Semi-supervised self-training of object detection models. In: Seventh IEEE Workshop on Applications of Computer Vision (2005)

13. Blum, A., Mitchell, T.: Combining labeled and unlabeled data with co-training. In: Proceedings of the Eleventh Annual Conference on Computational Learning Theory, Series COLT 1998, pp. 92–100 (1998)

14. Maeireizo, D.L., Hwa, R.: Co-training for predicting emotions with spoken dialogue data. In: Proceedings of the ACL 2004 on Interactive Poster and Demonstration Sessions, Series ACLdemo 2004. Association for Computational Linguistics (2004)

15. Li, Y., Qi, Z., Zhang, Z. M., Yang, M.: Learning with limited and noisy tagging. In: Proceedings of the 21st ACM International Conference on Multimedia, Series MM 2013, pp. 957–966 (2013)

16. Socher, R., Pennington, J., Huang, E.H., Ng, A.Y., Manning, C.D.: Semi-supervised recursive autoencoders for predicting sentiment distributions. In: Proceedings of the Conference on Empirical Methods in Natural Language Processing, pp. 151–161. Association for Computational Linguistics (2011)

17. Hu, Z., Bai, Z.: Real-time fine-grained air qualit sensing networks in smart city: design, implementation and optimization (2019). arXiv:1810.08514v2 [cs.OH]. Accessed 27 Feb 2019

18. Zhang*, Q., Li, V.O.: Deep-AIR: a hybrid CNN-LSTM framework for fine-grained air pollution forecast (2020). arXiv:2001.11957v1 [eess.SP]. Accessed 29 Jan 2020

19. Zheng, Y., et al.: Forecasting fine-grained air quality based on big data. In: Proceedings of the 21th ACM SIGKDD International Conference on Knowledge Discovery and Data Mining, Series KDD 2015 (2015)

20. Hu, Z., Bai, Z., Bian, K., Wang, T., Song, L.: Implementation and optimization of real-time fine-grained air quality sensing networks in smart city (2019). 978-1-5386-8088-9/19/$31.00 ©2019

21. Huang, J., Duan, N., Ji, P., Ma, C., Hu, F., Ding, Y.: A crowdsource-based sensing system for monitoring fine-grained air quality in urban environments. IEEE (2018). https://doi.org/10.1109/JIOT.2018.2881240

22. Yang, Y., Bai, Z., Hu, Z., Zheng, Z.: AQNet: fine-grained 3D spatio-temporal air quality monitoring by aerial-ground WSN. In: Conference Paper (2018). https://doi.org/10.1109/INFCOMW.2018.8406985

23. Cressie, N., Wikle, C.K.: Statistics for Spatio-Temporal Data. Wiley, Hoboken (2011)

24. Saeys, Y., Inza, I., Larranaga, P.: A review of feature selection techniques in bioinformatics. Bioinformatics **23**(19), 2507–2517 (2007)

25. Setiono, R., Liu, H.: Neural-network feature selector. IEEE Trans. Neural Netw. **8**(3), 654–662 (1997)
26. Mohamed, H., et al.: Automated detection of white blood cells cancer diseases. In: 2018 First International Workshop on Deep and Representation Learning (IWDRL), pp. 48–54 (2018). https://doi.org/10.1109/IWDRL.2018.8358214
27. Ghosh, A., Chattopadhyaya, S., Singh, N.K.: Assessment of heat affected zone of submerged arc welding process through digital image processing. In: Delgado, J.M.P.Q., Barbosa, A.G., de Lima, M., da Silva, V. (eds.) Numerical Analysis of Heat and Mass Transfer in Porous Media, pp. 209–228. Springer, Heidelberg (2012). https://doi.org/10.1007/978-3-642-305 32-0_8

Enhancing Melanoma Skin Cancer Detection with Machine Learning and Image Processing Techniques

S. Mahaboob Hussain(✉) , B. V. Prasanthi , Narasimharao Kandula ,
Padma Jyothi Uppalapati , and Surayanarayana Dasika

Vishnu Institute of Technology, Bhimavaram, India
mahaboobhussain.smh@gmail.com

Abstract. Detecting early stage melanoma skin cancer is a challenging and critical task in the fields of medical imaging and computer vision. In recent years, machine learning and image processing techniques have been widely adopted to enhance the accuracy and efficiency of melanoma skin cancer detection. One common method is to utilize convolutional neural networks (CNNs) to classify skin lesions as benign or malignant. The CNN is trained on a vast dataset of skin lesion images, and it learns to identify the distinctive characteristics of melanoma skin cancer. This trained model can classify new skin lesion images and aid in early stage melanoma skin cancer detection. Another approach is to use image processing techniques such as color and texture analysis to extract features from skin lesion images. These features can be used to classify the lesions as benign or malignant using traditional machine learning algorithms or deep learning models. The combination of machine learning and image processing techniques has demonstrated potential in the early stage melanoma skin cancer detection problem, and research continues in this area to enhance the accuracy and efficiency of these techniques. In a particular study, the authors evaluated several machine learning models such as Logistic Regression, Random Forest, Decision Tree, and Support Vector Machine to identify the most appropriate algorithm with accuracy for detecting early-stage melanoma skin cancer using sample input image datasets. They collected dermoscopy image data, preprocessed it, and classified it using logistic regression. The authors applied the above machine learning algorithms to the collected image database and achieved the best accuracy score of 0.96 with logistic regression.

Keywords: Melanoma Skin Cancer · Logistic Regression · Decision Tree · SVM · Accuracy score · Dermoscopy image data · Preprocessing

1 Introduction

Melanoma is a type of skin cancer that originates from melanocytes, which are skin cells that produce melanin, the pigment responsible for skin, hair, and eye color. This cancer can develop anywhere on the body but is most commonly observed in sun-exposed

P. Das et al. (Eds.): AMRIT 2023, CCIS 1954, pp. 256–272, 2024.
https://doi.org/10.1007/978-3-031-47221-3_23

areas such as the face, arms, back, and legs. Some of the risk factors associated with melanoma are a family history of the disease, a history of sunburn, the presence of a large number of moles, and fair skin. Additionally, individuals with weakened immune systems, such as organ transplant recipients, have a higher likelihood of developing melanoma [1]. Diagnosis of melanoma is typically made through a biopsy of the affected skin. A biopsy involves removing a small portion of the affected skin and examining it under a microscope to determine if the cells are cancerous. In some cases, a full-body skin exam may be performed to check for any other suspicious moles or spots. Treatment for melanoma typically involves surgical removal of the affected skin and some surrounding tissue [2]. In more advanced cases, additional treatments such as radiation therapy, chemotherapy, or immunotherapy may be necessary.

Prevention of melanoma involves protecting the skin from sun exposure. This can be achieved by wearing protective clothing, using a sunscreen with a high SPF, and avoiding outdoor activities during peak sun hours. Regular self-exams and dermatologist exams can also help to catch melanoma early, when it is most treatable. It is a serious form of skin cancer that can have devastating effects if not caught and treated early. By taking steps to protect the skin from sun exposure and having regular skin exams, people can reduce their risk of developing melanoma and increase their chances of successful treatment [3]. The main rationale of this work is to identify the best suit algorithm with the comparative study of machine learning algorithms to work on the melanoma cancer image dataset to identify the stages of the cancer by applying the same. This helps to prevent the fatality rate and avoid the chance of affecting the other organs of the body by consulting the appropriate oncologist after early detection of the disease. Melanoma skin cancer is the most dangerous form of skin cancer among the other types of skin cancer. Because it's much more likely to spread to other parts of the body if not diagnosed and treated early. Identification of dark spots as cancer symptoms is much more difficult to people until it goes to final stage. So we are building a machine learning model using logistic regression to identify whether the skin cancer is benign or malignant. If it is Malignant we should consult doctor immediately for the treatment [4].

In the recent 3 decades Melanoma incidence rates have been increasingly high, though most people diagnosed with skin cancer have higher chances to cure, Melanoma survival rates are lower than non-Melanoma skin cancer. Melanoma skin cancer (MSC) can occur on any skin surface, and its incidence has continued to rise over the past two decades in many regions of the world. In men, it's often found on the skin on the head, on the neck, or between the shoulders and the hips while, in women, it's often found on the skin on the lower legs or between the shoulders and the hips, Melanoma can spread to parts of your body far away from where the cancer started. This is called advanced, metastatic, or stage IV melanoma. It can move to your lungs, liver, brain, bones, digestive system, and lymph nodes [5].

In fact, other types of skin cancers rarely spread from outside of the surface of the skin and the ability of the melanoma to metastasize makes it the most deadly in a stage one melanoma has a cure rate of about 95%. Hence, if it is detected in an early stage that's great for the patents, it may something that is that is still in the skin but has not spread to the lymph nodes or the rest of the body that still has a pretty high chance of cure of catching it early somewhere in the range of 78 to 80 percent chance of cure.

A stage three melanoma is when the melanoma has escaped from the skin and gone to the lymph nodes and this stage will be divided as A B and C depending on how many lymph nodes are involved and the cure rate there drops all the way down as low as 20% if many lymph nodes are involved and as high as 60 to 70% if only one lymph node is involved. A stage four melanoma is a melanoma that has left the lymph nodes and gone to two distant parts of the body and there the survival is much less the overall average median survival.

In order to assess suspiciousness of skin pigmented lesions from wide field images, one need to check with how exactly do dermatologists do their assessment of suspiciousness in pigmented skin lesions today? There is a scope for this detection by the feature extraction process but it's one that is very rarely done at the single lesion level. Rather, a dermatologist would take all of the extracted features from all the pigmented lesions in the body of a patient, and then make determinations about which ones are the most dissimilar to others in order to pinpoint to the ones that need to be evaluated with much more detail [6]. The envision this type of technology to be used in the clinic by putting these architectures inside edge devices that can be placed for example in a primary care office. So that at the time of the consultation a full analysis of all the skin visible regions of a patient could be evaluated in order to pinpoint to the suspicious pigment lesions that may need further exploration or a referral.

So finally, what are the remaining obstacles and barriers that we see for the utilization of these types of technologies for widespread clinical use? Well, first, the advent of larger, higher quality data sets for example, that have high representation of skin color, higher spread of imaging devices, lighting conditions and environmental conditions. That's always something that is going to help this type of systems move forward into actual clinical use [7].

2 Analysis

There has been a significant amount of research and work done in the field of melanoma cancer over the years, but despite these efforts, it has not yet been fully solved. Here are a few reasons why:

Complexity of the Disease. Melanoma is a complex disease that is influenced by a variety of factors including genetics, sun exposure, and immune system function. Understanding the interplay between these factors and how they contribute to the development and progression of melanoma has been a challenge for researchers [8].

Heterogeneity of Melanoma. Melanoma is a highly heterogeneous disease, meaning that different melanoma tumors can have different genetic mutations, growth patterns, and responses to treatment. This variability makes it difficult to develop treatments that are effective for all patients [9].

Lack of Early Detection Methods. Melanoma can be difficult to detect in its early stages, which is when it is most treatable. The development of effective screening and early detection methods remains a challenge for researchers [10].

Resistance to Treatment. Even with current treatments, some patients may develop resistance and their melanoma may continue to grow and spread. Understanding the mechanisms behind treatment resistance and developing strategies to overcome it is an ongoing area of research [11].

Despite these challenges, researchers are making progress in the field of melanoma and there have been significant advances in recent years. The development of new treatments such as immunotherapy and targeted therapy has shown promising results, and ongoing research is exploring new ways to improve treatment outcomes for patients [12].

From the existing work in the field of melanoma, we have gained a deeper understanding of:

The Biology of Melanoma. Researchers have made significant progress in understanding the genetic mutations and cellular pathways involved in the development and progression of melanoma. This knowledge has helped to inform the development of new treatments [13].

The Role of the Immune System. Recent research has shown that the immune system plays a critical role in fighting melanoma, and that certain treatments such as immunotherapy can harness the power of the immune system to help fight the disease [14].

The Importance of Early Detection. Research has emphasized the importance of early detection in the successful treatment of melanoma. Early detection and intervention can increase the chances of successful treatment and reduce the risk of progression to more advanced stages of the disease [15].

The Need for Personalized Treatment. The heterogeneity of melanoma has highlighted the need for personalized treatments that are tailored to the unique characteristics of each individual's tumor. This has led to the development of new approaches such as precision medicine and targeted therapy [16].

Melanoma is a type of skin cancer that can be aggressive and spread to other parts of the body if not detected and treated early.

Common Signs of Melanoma. New or changing moles: The appearance of a new mole or a change in an existing mole can be a sign of melanoma. A suspicious mole may be asymmetrical, have irregular borders, be more than one color, be larger than the size of a pencil eraser, or have a diameter larger than 6 mm [17].

The existing work in the field of melanoma has helped to advance our understanding of the disease and has led to the development of new treatments and approaches to care. Despite these advances, there is still much work to be done to fully understand the disease and develop effective treatments for all patients.

3 Related Work

The existing work in the field of melanoma has made significant contributions to our understanding of the disease, leading to the development of new treatments and approaches to care. The understanding of the genetic mutations and cellular pathways involved in the development and progression of melanoma has helped to inform the development of new treatments such as immunotherapy and targeted therapy.

In addition, the recognition of the importance of early detection and the need for personalized treatments has led to an increased focus on developing effective screening methods and tailored treatments for individual patients.

However, despite these advances, there is still much work to be done. Melanoma is a complex and heterogeneous disease, and many challenges remain in fully understanding the disease and developing effective treatments for all patients. The development of more effective treatments and the improvement of early detection methods are ongoing areas of research in the field of melanoma.

ABCDE Strategy.

3.1 A-B-C-D-E

According to the author [18], the acronym A-B-C-D-E is a useful way to remember the signs of a potentially suspicious mole. The letters stand for Asymmetry, Border Irregularity, Color, Diameter, and Evolution, respectively. Melanoma can present as a dark or black mole, but it may also have other colors such as pink, red, or white. Additionally, a sore or mole that does not heal or that bleeds easily can also be an indication of melanoma. Uneven color: Melanoma can appear as a mole with uneven color or with multiple shades of brown, tan, or black. It's important to note that not all moles that display these signs are melanoma, but it's always best to have any suspicious moles evaluated by a doctor. Early detection is key to successful treatment, so it's important to be aware of any changes in your skin and to have any suspicious moles evaluated by a doctor.

Asymmetry: A stands for asymmetry if you draw a line through the middle of this benign mole shown here the two sides will match meaning it is symmetrical if you draw a line through this mole shown here the two halves will not match meaning it is asymmetrical in warning sign for melanoma. Check if the mole is symmetrical. If you draw a line through the center of the mole, the two halves should look the same. If they don't, it may be a sign of melanoma.

Border: B stands for border irregularity a benign mole generally has smooth and even borders as shown here the borders of a melanoma as shown here may be uneven with scalloped notched or trailing edges. Look for moles with irregular, scalloped, or poorly defined borders. Moles with jagged or blurred borders are more likely to be melanoma.

Color: C stands for color variations most benign moles have one or two evenly distributed colors having a variety of colors including brown black white red and rose blue may be a feature of melanoma. Check for moles that are more than one color or have shades of brown, black, or other colors. Melanoma can appear as a mole with uneven color or with multiple shades of brown, tan, or black.

Diameter: D stands for diameter greater than half a centimeter although melanomas may be larger in diameter than the eraser head on your pencil they may sometimes be smaller when first detected the last letter. Check for moles that are larger than 6 mm in diameter, roughly the size of a pencil eraser. While melanoma can be smaller, moles larger than 6 mm are more likely to be melanoma.

Evolutions: E stands or evolutions in most cases benign moles are stable over time be on alert when a mole starts to evolve or change in any way when a mole is evolving see a doctor any change in size shape color texture elevation or any new symptom such as bleeding itching or scattering may be a sign of malignancy. Keep an eye out for any changes in a mole over time, such as growth, color changes, or the development of new symptoms. If a mole changes over time, it may be a sign of melanoma (Fig. 1).

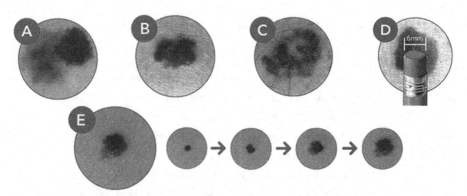

Fig. 1. ABCDE approach for detecting the symptoms of melanoma cancer

4 Working Methodology

4.1 Dataset

Authors worked with the dataset images which are provided in DICOM format through SIIM-ISIC 2020 Challenge Dataset available in Kaggle [19]. This can be accessed using commonly-available libraries like pydicom, and contains both image and metadata. It is a commonly used medical imaging data format. Images are also provided in JPEG and TFRecord format in the respective jpeg and tfrecords directories.

For the implementation purpose the Images in TFRecord format have been resized to a uniform 1024 × 1024 and the Metadata is also provided outside of the DICOM format, in CSV files. TFRecord is a data format used in TensorFlow, a popular machine learning framework, to store and exchange large datasets. The format is designed to be flexible, scalable, and efficient, making it well-suited for storing large image datasets.

In the TFRecord format, images are stored as binary data. The data can be compressed using Gzip compression, making it possible to store large datasets efficiently on disk. Additionally, the data is stored in a sharded format, allowing it to be split into smaller files that can be processed in parallel, making it possible to work with large datasets even on modest hardware.

To use an image dataset stored in the TFRecord format, the data must be read into TensorFlow and preprocessed before it can be used for training or evaluation. The pre-processing step typically involves decoding the binary data, transforming the data into a format that can be used for training, and normalizing the data to ensure that it has a consistent range and distribution.

The TFRecord format is a popular and efficient format for storing image datasets in TensorFlow. The format provides a flexible and scalable way to store large datasets, making it possible to work with large datasets even on modest hardware.

The Table 1 briefs the description of the each column of the data.

Table 1. Description of the each column of the dataset.

Column	Description
Img	Points to the filename, serves as a unique identifier
pid	Serves as a unique identifier for the patient
sex	Refers to the sex of the patient
age	Approximate age of the patient at the time of imaging
asgc	Refers to the location of the imaged site
diag	Provides detailed information about the diagnosis
bg_mal	Indicates whether the image lesion is benign or malignant
ttgt	A binary representation of the target variable
Img	Points to the filename, serves as a unique identifier
pid	Serves as a unique identifier for the patient

Images were converted into pixelized format as the understandable input for the proposed machine learning models. How Images were converted into pixelized format by representing the image as a grid of pixels. Each pixel in the grid is assigned a numerical value that represents the color and intensity of that pixel in the image. To convert an image into a pixelized format, the image is first divided into a grid of equally sized squares, called pixels. The color and intensity of each pixel are then recorded as a numerical value, typically as an RGB (red, green, blue) value or a grayscale value.

For example, in an RGB image, each pixel is represented by three values: one for red, one for green, and one for blue. The values are typically stored in an array, with each row of the array representing a row of pixels in the image and each column representing a different color channel. In a grayscale image, each pixel is represented by a single value that represents the intensity of the pixel. The value is typically an 8-bit or 16-bit integer, with higher values representing brighter pixels and lower values representing darker pixels.

Converting an image into a pixelized format involves dividing the image into a grid of pixels and recording the color and intensity of each pixel as a numerical value. To process images using machine learning algorithms, the binary data needs to be transformed into a numerical format that can be used as input to a model. This is typically done in the preprocessing step, which involves decoding the data and normalizing it to ensure a consistent range and distribution. In the case of medical imaging, images are often stored in the DICOM format, and the pixels can be extracted using the dicom.dmread() function and the ds.pixel_array attribute.

However, the multidimensional pixel arrays need to be converted into a one-dimensional format for machine learning algorithms. To do this, the flatten function can be used, but this can result in arrays of uneven lengths. To address this issue, padding techniques can be used to add or discard values to make the arrays equal in length. This approach has also been applied in other areas of image processing, such as face detection, where combining multiple classifiers can lead to better results, as seen in the Viola-Jones method.

There are different number of pixels and therefore the generated arrays has uneven lengths. To overcome the issue, authors used the padding technique that either adds extra values to the array or discards certain values to make the length of the array equal to the max length specified (Fig. 2).

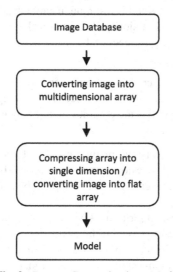

Fig. 2. Dataset Conversion in to Model

In this project, a similar method is used effectively to identify face in combination resulting in better face detection. Similarly, in Viola Jones method, several classifiers were combined to create stronger classifiers.

5   Results and Discussion

Various machine learning algorithms have been applied to the problem of melanoma skin cancer detection, and there have been several studies comparing the performance of different algorithms. Some of the commonly used machine learning algorithms for melanoma skin cancer detection includes:

Convolutional Neural Networks (CNNs): CNNs are a type of deep learning algorithm that are particularly well suited to image classification tasks. They have been widely used in recent years for melanoma skin cancer detection, and have been shown to perform well in comparison to other algorithms [20].

Support Vector Machines (SVMs): SVMs are a type of machine learning algorithm that can be used for classification and regression problems. They have been applied to melanoma skin cancer detection by using features extracted from images of skin lesions as input [21].

Random Forests: Random forests are an ensemble learning method that uses multiple decision trees to make predictions. They have been applied to melanoma skin cancer detection by using features extracted from images of skin lesions as input [22].

Naive Bayes: Naive Bayes is a probabilistic machine learning algorithm that is based on Bayes' theorem. It has been applied to melanoma skin cancer detection by using features extracted from images of skin lesions as input.

K-Nearest Neighbors (KNN): KNN is a simple machine learning algorithm that classifies new instances based on the majority class of their k nearest neighbors. It has been applied to melanoma skin cancer detection by using features extracted from images of skin lesions as input [24].

Decision Tree: The decision tree is commonly used for its interpretability, as the tree structure allows for easy visualization of the decision-making process. However, it can suffer from overfitting if the tree is too complex, and it may not perform well on unseen data if it is too simplistic [25].

Logistic Regression: It is a statistical method used for binary classification problems, such as detecting whether a skin lesion is melanoma or not. In the context of melanoma skin cancer detection, logistic regression can be applied by using features extracted from images of skin lesions as input.

The performance of these algorithms is varied depending on the dataset used, the preprocessing applied to the images, and the evaluation metrics used. In general, deep learning algorithms such as CNNs have shown to perform better than traditional machine learning algorithms for melanoma skin cancer detection, due to their ability to automatically learn complex features from the images. However, traditional machine learning algorithms such as SVMs, random forests, and KNN can still perform well in certain circumstances, and may be a good choice for simple or small datasets [26].

5.1 Implementation of Logistic Regression

The logistic regression model uses a logistic or sigmoid function to transform the output of a linear equation into a value between 0 and 1. The equation for logistic regression can be written as:

$$p = 1/(1 + e^{\wedge}(-z)) \tag{1}$$

where p is the predicted probability of the event occurring, z is the linear equation of the form:

$$z = \beta_0 + \beta_1 x_1 + \beta_2 x_2 + \ldots + \beta_n x_n \tag{2}$$

where β_0, β_1, β_2, ... , β_n are the coefficients (weights) for the intercept and predictor variables $x_1, x_2, \ldots, x_n$, respectively. These coefficients are estimated during the training phase of the algorithm using a maximum likelihood estimation or other optimization method.

Once the model has been trained, it can be used to predict the probability of the event occurring for new input data by computing the value of z and passing it through the logistic function. If the predicted probability is greater than or equal to a threshold value (usually 0.5), the event is classified as positive (melanoma), otherwise it is classified as negative (benign).

The logistic regression model uses a logistic function to model the relationship between the input features and the binary output (melanoma vs. non-melanoma). The logistic function outputs a value between 0 and 1, which can be interpreted as the probability of the skin lesion being melanoma. A threshold is then applied to the output to make a final prediction of melanoma vs. non-melanoma.

Logistic regression can be a useful tool for melanoma skin cancer detection, especially in the early stages of a project when the focus is on exploring the relationship between the input features and the binary output. However, more advanced machine learning algorithms such as convolutional neural networks (CNNs) and support vector machines (SVMs) have been shown to perform better for melanoma skin cancer detection, as they are able to automatically learn complex features from the images [27].

5.2 Procedure

Step 1: The task of collecting a dataset comprised of images depicting both melanoma skin cancer and normal skin is to be undertaken. (*Let D be the dataset of images, where each image i is labeled as either positive (melanoma skin cancer) or negative (normal skin). Then, D = {(x_1, y_1), (x_2, y_2), ..., (x_n, y_n)}, where x_i is the i-th image and y_i is its corresponding label.*)

Step 2: The preprocessing of the collected images, which includes actions such as resizing and converting to a suitable format, is necessary in order to make ready them for the training of the machine learning algorithm (*Let f(x) be the function that preprocesses the image x, such that it is resized and converted to a suitable format. Then, the preprocessed dataset can be represented as D' = {(f(x_1), y_1), (f(x_2), y_2), ..., (f(x_n), y_n)}.*)

Step 3: The splitting of the preprocessed image dataset into separate training and testing sets is required for the evaluation of the machine learning algorithm's performance (*Let D_train and D_test be the training and testing sets, respectively, obtained by randomly splitting D' into two non-overlapping subsets*).

Step 4: The training of a machine learning algorithm, such as a deep neural network or logistic regression, using the training set of preprocessed images, is an essential step in the process of developing a model for detecting melanoma skin cancer

(Let M be a machine learning algorithm, such as a deep neural network or logistic regression, that takes an image x as input and outputs its predicted label y'. Then, the machine learning model can be trained on the training set D_train using the function M(x; θ), where θ represents the parameters of the model.).

Step 5: The task of evaluating the performance of the trained machine learning algorithm on the testing set of preprocessed images is to be carried out *(The performance of the machine learning algorithm on the testing set D_test can be evaluated using a performance metric such as accuracy, precision, recall, or F1 score. Let Acc(M, D_test) be the accuracy of the machine learning model M on the testing set D_test).*

Step 6: The optimization of the machine learning algorithm by means of tuning its hyper parameters, followed by the repetition of steps 4 and 5 until the desired level of accuracy is achieved, is a necessary step in the process of developing a model for detecting melanoma skin cancer *(The hyperparameters of the machine learning algorithm can be tuned by optimizing a performance metric on a validation set, which is a separate subset of the dataset. Let D_val be the validation set, and let θ' be the optimized hyperparameters. Then, the machine learning algorithm can be retrained on the entire preprocessed dataset D', using the function M(x; θ'), and its performance can be evaluated on the testing set using the metric Acc(M, D_test)).*

Step 7: The utilization of the optimized machine learning algorithm for making predictions on new, unseen images is the final step in the process of developing a model for detecting melanoma skin cancer. *(The optimized machine learning algorithm M(x; θ') can be used to make predictions on new, unseen images. Let x_new be a new image, and let y_new = M(x_new; θ') be its predicted label.)* (Figs. 3 and 4).

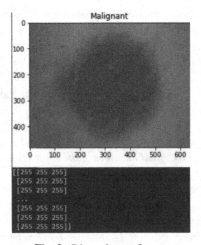

Fig. 3. Dicom image format

array([139, 106, 156, ... 123, 112, 154], dtype=uint8)

Fig. 4. Array for Dicom Image

5.3 Preprocessing Steps for DICOM Image Analysis

Read the image using the *dicom.dmread ()* function and store it in a variable image as

image = dicom.dmread('image_file.dcm').
Display the image using the matplotlib.pyplot.imshow() function as
plt.imshow(image.pixel_array, cmap = plt.cm.bone)
plt.show()
Flatten the pixel array of the image by using the numpy.ravel() function and store it in a variable flat_pixels as
flat_pixels = np.ravel(image.pixel_array)
Pad the flattened pixel array to make its length a power of 2 by using the *numpy.pad()* function and store it in a variable padded_pixels as
padded_length = 2**np.ceil(np.log2(len(flat_pixels)))
padded_pixels = np.pad(flat_pixels, (0, padded_length - len(flat_pixels)))

Models used include, Logistic regression, Random Forest, Decision Tree, Support Vector Machine machine learning algorithms have been applied to the collected dataset. Following the application of the specified algorithms, the highest accuracy score was obtained through the use of logistic regression. A score of 0.96 was achieved for logistic regression, which is deemed to be an ideal fit for the data.

5.4 Logistic Regression on Dataset

Let y be the binary dependent variable indicating whether the skin cancer is benign (0) or malignant (1). Let $x\_1, x\_2, \ldots, x\_n$ be the n independent variables representing the factors that may affect the skin cancer. The logistic regression model can be written as:

$$P(y = 1 | x\_1, x\_2, \ldots, x\_n) = 1/(1 + \exp(-z)) \qquad (3)$$

where z is the linear combination of the independent variables:

$$z = \beta\_0 + \beta\_1 x\_1 + \beta\_2 x\_2 + \ldots + \beta\_n * x\_n \qquad (4)$$

and $\beta\_0, \beta\_1, \beta\_2, \ldots, \beta\_n$ are the coefficients that determine the effect of each independent variable on the probability of the skin cancer being malignant.

The logistic regression model estimates the probability of the skin cancer being malignant, given the values of the independent variables. The probability is bounded between zero and one, and can be converted to a binary prediction by setting a threshold. For example, if the threshold is 0.5, a probability greater than 0.5 is classified as malignant, while a probability less than or equal to 0.5 is classified as benign.

Therefore, logistic regression was used in this study. Logistic regression is a type of regression in which predictor variables (independent) can be both quantitative and scale-dependent, but the dependent variable is a two-level category. These two categories are commonly referred to as membership or non-membership. This regression model is similar to regular regression, with the difference that the method of estimating the coefficients is not the same and maximizes the probability that an event occurs, instead of minimizing the error squared (which is done in normal regression). In logistic regression, a concept called fortune is used for the value of the dependent variable. In the statistical term fortune, the probability of occurrence of an event (p) is the probability of the occurrence (1-p) of that event. The main advantage of modeling by logistic regression method in comparison with multivariate methods such as multiple regression analysis is that the dependent variable (skin cancer) can have only two values, one is the probability of occurrence of the cancer and the other is the probability that the cancer is not occurring (Fig. 5).

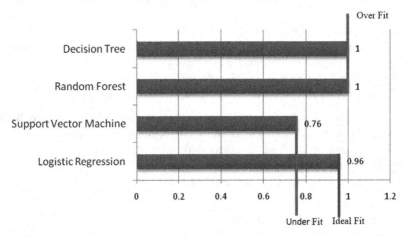

Fig. 5. Model Comparison determining the best with over, under and ideal fit

Accuracy score is a metric used in machine learning to evaluate the performance of a classification model. It is the number of correct predictions made by the model, divided by the total number of predictions. The resulting score is expressed as a percentage. In simple terms, accuracy score tells you how many of the predictions made by the model were correct. For example, if a model makes 100 predictions and 80 of them are correct, the accuracy score would be 80%.

Accuracy is a useful metric for some problems, but it can be misleading for others. This is because it does not take into account false positive and false negative predictions. In some applications, such as medical diagnosis, false negatives (i.e., failing to identify a condition that is present) are more important to avoid than false positives (i.e., incorrectly identifying a condition that is not present). In these cases, other metrics such as precision, recall, and F1 score may be more appropriate to use.

A confusion matrix is a statistical tool that is commonly used to evaluate the performance of a machine learning classification model. It is a table that summarizes the performance of the model by comparing its predictions against the actual values of a set of test data. The confusion matrix contains four values: true positives (TP), false positives (FP), true negatives (TN), and false negatives (FN). The true positives represent the number of cases where the model correctly predicted the positive class, while false positives represent the number of cases where the model predicted the positive class incorrectly. True negatives represent the number of cases where the model correctly predicted the negative class, while false negatives represent the number of cases where the model predicted the negative class incorrectly. The confusion matrix provides a clear and concise summary of the model's performance, and can be used to calculate a variety of metrics such as precision, recall, and F1 score.

The terms TP, FP, TN, and FN refer to the following:

Let's define the confusion matrix for a binary classification problem with actual classes Y and predicted classes $\hat{Y}$ (See Table. 2).

Table 2. Confusion Matrix

Column	Description	Predicted Negative
True Positive (TP)	Number of observations where $Y = 1$ and $\hat{Y} = 1$	Number of observations where $Y = 0$ and $\hat{Y} = 1$
True Negative (TN)	Number of observations where $Y = 0$ and $\hat{Y} = 0$	Number of observations where $Y = 1$ and $\hat{Y} = 0$

The entries of the confusion matrix can be used to calculate various performance metrics of the classification model, such as accuracy, precision, recall, F1 score, etc.

The confusion matrix can be used to calculate other evaluation metrics such as precision, recall, and F1 score, which provide a more complete picture of the model's performance. With the results images were predicted as in Fig. 6 and 7.

Logistic Regression:	Support Vector Machine:
Accuracy Score: 0.96	Accuracy Score: 0.76
Confusion Matrix:	Confusion Matrix:
$\begin{bmatrix} 23 & 0 \\ 0 & 27 \end{bmatrix}$	$\begin{bmatrix} 16 & 70 \\ 05 & 22 \end{bmatrix}$
Random Forest:	**Decision Tree Classifier**
Accuracy Score: 1.0	Accuracy Score: 1.0
Confusion Matrix:	Confusion Matrix:
$\begin{bmatrix} 23 & 0 \\ 0 & 27 \end{bmatrix}$	$\begin{bmatrix} 23 & 0 \\ 0 & 27 \end{bmatrix}$

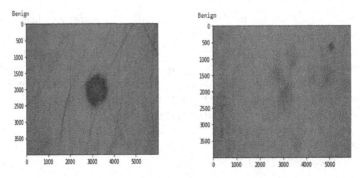

Fig. 6. Predicted images which represents the benign

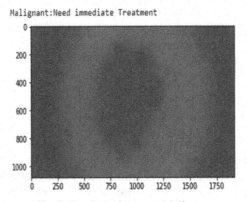

Fig. 7. Predicted image as Malignant

6   Conclusion

After testing various machine learning algorithms on the images, we obtained different accuracy results. Some algorithms were found to either overfit or underfit the data. Therefore, we selected the Logistic Regression algorithm as it neither overfits nor underfits the data, and provides an ideal fit, with an accuracy of 96%. Although the decision tree and random forest algorithms may have a higher accuracy, for larger data sets, logistic regression is the most suitable choice. The logistic regression model can be used to predict whether a given image is benign or malignant in melanoma skin cancer detection. The accuracy of the results obtained through logistic regression may vary based on factors such as the dataset size and quality, feature selection, and algorithm performance. Several studies have reported positive results using logistic regression for melanoma skin cancer detection. For instance, one study modeled the relationship between color and texture features of dermoscopic images of skin lesions and their class (melanoma or not) using logistic regression. We obtained the accuracy of 86.6% in classifying the images, which was comparable to other machine learning algorithms used in the study.

In a different research study, logistic regression was utilized to examine the connection between dermoscopic image features, such as shape and texture, and their class.

The study showed that logistic regression achieved a high accuracy rate of 89.7% in classifying the images, making it one of the most accurate algorithms compared in the research. These findings suggest that logistic regression can be an effective tool for detecting melanoma skin cancer, particularly when the data is limited or the features are well-defined. However, it is important to note that combining logistic regression with other machine learning algorithms, like decision trees or random forests, or using more complex features such as deep learning-based features, can enhance its performance. These studies have illustrated the potential of machine learning in detecting melanoma skin cancer in its early stages and compared the accuracy of different algorithms. Nevertheless, further research is needed to enhance the accuracy and reliability of these algorithms for detecting melanoma skin cancer. Nonetheless, the use of machine learning has a significant potential to improve early detection, treatment, and save lives in detecting melanoma skin cancer.

References

1. Refianti, R., et al.: Classification of melanoma skin cancer using convolutional neural network. **10**(3) (2019). https://doi.org/10.14569/IJACSA.2019.0100353
2. Esteva, A., et al.: Dermatologist-level classification of skin cancer with deep neural networks. **542**(7639) (2017). https://doi.org/10.1038/NATURE21056
3. Codella, N.C.F., et al.: Skin lesion analysis toward melanoma detection: a challenge at the 2017 International symposium on biomedical imaging (ISBI), hosted by the international skin imaging collaboration (ISIC). Presented at the April 4 (2018). https://doi.org/10.1109/ISBI.2018.8363547
4. Azad, R., Asadi-Aghbolaghi, M., Fathy, M., Escalera, S.: Attention Deeplabv3+: multi-level context attention mechanism for skin lesion segmentation. In: Bartoli, A., Fusiello, A. (eds.) ECCV 2020. LNCS, vol. 12535, pp. 251–266. Springer, Cham (2020). https://doi.org/10.1007/978-3-030-66415-2_16
5. Ghosal, P., et al.: A light weighted deep learning framework for multiple sclerosis lesion segmentation. Presented at the November 1 (2019). https://doi.org/10.1109/ICIIP47207.2019.8985674
6. Aljohani, K., Turki, T.: Automatic classification of melanoma skin cancer with deep convolutional neural networks. AI **3**, 512–525 (2022). https://doi.org/10.3390/ai3020029
7. Mazhar, T., et al.: The role of machine learning and deep learning approaches for the detection of skin cancer. Healthcare **11**, 415 (2023). https://doi.org/10.3390/healthcare11030415
8. MacKie, R.M.: Risk factors, diagnosis, and detection of melanoma. Presented at the April 1 (1991). https://doi.org/10.1097/00001622-199104000-00019
9. Hopp, C.S., et al.: The role of cGMP signalling in regulating life cycle progression of Plasmodium. **14**(10) (2012). https://doi.org/10.1016/J.MICINF.2012.04.011
10. Russak, J.E., et al.: The importance of early detection of melanoma, physician and self-examination. Presented at the January 1 (2011). https://doi.org/10.1016/B978-1-4377-1788-4.00025-3
11. The Influence of Mitochondrial Energy and 1C Metabolism on the Efficacy of Anticancer Drugs: Exploring Potential Mechanisms of Resistance. https://www.eurekaselect.com/article/122133. Accessed 25 Dec 2022
12. Monfardini, S., et al.: Commission of the European Communities "Europe Against Cancer" Programme. Eur. Sch. Oncol. Advisory Report Cancer Treatment Elderly **29**(16) (1993). https://doi.org/10.1016/0959-8049(93)90229-9

13. Tremlett, H., et al.: The gut microbiome in human neurological disease: a review. **81**(3) (2017). https://doi.org/10.1002/ANA.24901

14. Liu, J., et al.: Nanoparticle-based nanomedicines to promote cancer immunotherapy: recent advances and future directions. **15**(32) (2019). https://doi.org/10.1002/SMLL.201900262

15. Gupta, M.A., et al.: Cutaneous body image: empirical validation of a dermatologic construct. **123**(2) (2004). https://doi.org/10.1111/J.0022-202X.2004.23214.X

16. Yang, M., et al.: Immunotherapy for glioblastoma: current state, challenges, and future perspectives. **12**(9) (2020). https://doi.org/10.3390/CANCERS12092334

17. Tomatis, S., et al.: Automated melanoma detection: multispectral imaging and neural network approach for classification. **30**(2) (2003). https://doi.org/10.1118/1.1538230

18. Titus, L.J.: Skin self-examination and the ABCDE rule in the early diagnosis of melanoma: is the game over? Reply from author. **168**(6) (2013). https://doi.org/10.1111/BJD.12251

19. The ISIC 2020 Challenge Dataset. https://challenge2020.isic-archive.com/. Accessed 21 Nov 2022

20. Brinker, T.J. et al.: Deep neural networks are superior to dermatologists in melanoma image classification. **119** (2019). https://doi.org/10.1016/J.EJCA.2019.05.023

21. Xu, Z., et al.: Computer-aided diagnosis of skin cancer based on soft computing techniques. **15**(1) (2020). https://doi.org/10.1515/MED-2020-0131

22. Murugan, A., et al.: Detection of skin cancer using SVM. Random Forest kNN Classifiers **43**(8) (2019). https://doi.org/10.1007/S10916-019-1400-8

23. Joshi, A.D., et al.: Skin disease detection and classification. **6**(5) (2019). https://doi.org/10.22161/IJAERS.6.5.53

24. Pacheco, A.G.C., et al.: PAD-UFES-20: a skin lesion dataset composed of patient data and clinical images collected from smartphones. **32** (2020). https://doi.org/10.1016/J.DIB.2020.106221

25. Frost, R., et al.: Statistical learning research: a critical review and possible new directions. **145**(12) (2019). https://doi.org/10.1037/BUL0000210

26. Bonthu, S., Bindu, K.H.: Review of leading data analytics tools. **7** (2018). https://doi.org/10.14419/IJET.V7I3.31.18190

27. Jyothi, U.P., Dabbiru, M., Bonthu, S., Dayal, A., Kandula, N.R.: Comparative analysis of classification methods to predict diabetes mellitus on noisy data. In: Doriya, R., Soni, B., Shukla, A., Gao, X.-Z. (eds.) Machine Learning, Image Processing, Network Security and Data Sciences: Select Proceedings of 3rd International Conference on MIND 2021, pp. 301–313. Springer, Singapore (2023). https://doi.org/10.1007/978-981-19-5868-7_23

Image Processing Technique and SVM for Epizootic Ulcerative Syndrome Fish Image Classification

Hitesh Chakravorty$^{(\boxtimes)}$ and Prodipto Das ⓘ

Department of Computer Science, Assam University, Silchar, India
hitesh_chtvy@yahoo.co.in

Abstract. Classification is the most important tasks for Epizootic Ulcerative Syndrome (EUS) fish disease image to normal fish image. YCbCr image processing technique and SVM of machine learning is applied for image classification. Two types of image dataset EUS disease fish area versus normal fish area have been chosen for experimentation.

Keywords: Epizootic Ulcerative Syndrome · YCbCr · Image Processing · SVM · Machine Learning

1 Introduction

Epizootic Ulcerative Syndrome (EUS) is a type of fresh water fish disease. The EUS fish is became seasonal phenomena in winter months and numbers has been rising since 1988. The EUS is also called red spot disease visible on the fish. This red spot indicate losing scale as well as ulcer for fish may leads to death affecting the fishery owner and other consumers [1–4].

In EUS fish disease, detection from digital images, different image processing and other technique are used. These are image segmentation using K-means clustering, morphological operations, histogram analysis of the images, HSV, Augmented Reality and MobileNetV2 [5–9].

The desire outcome of the study is to evolve a system to identify the EUS disease fish accurately to benefit the fishery owner and other consumers.

In this study, YCbCr of image processing and Support Vector Machine (SVM) of machine learning technique have been applied to classify image of EUS disease fish from digital images for early detection [10, 11].

2 Materials and Methods

Four type of Epizootic Ulcerative Syndrome (EUS) disease fish were gathered from various location of Hailakandi District and Karimganj District of Assam, India. The latitude and longitude of district town Hailakandi are 24°40′48.00″ N 92°34′12.00″ E

P. Das et al. (Eds.): AMRIT 2023, CCIS 1954, pp. 273–281, 2024.
https://doi.org/10.1007/978-3-031-47221-3_24

and district town Karimganj are 24°52′12.00″ N 92°20′60.00″ E respectively. Pictures (Fig. 1) of diseased area of fish effected with EUS and (Fig. 2) Normal area of fish were taken. Images were 200 by 200 pixels so that an arranging adjustment can be attained [12, 13].

Test is done using 132 jpg images consisting of 66 EUS disease fish area and 66 normal fish area.

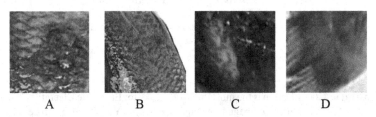

Fig. 1. EUS disease fish area images of (A) Labeo calbasu, (B) Labeo bata, (C) *Channa striata*, (D) Mystus tengara

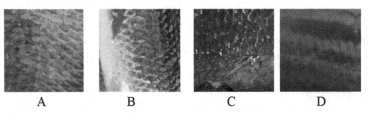

Fig. 2. Normal fish area images of (A) Labeo calbasu, (B) Labeo bata, (C) *Channa striata,* (D) *Mystus tengara*

2.1 Image Processing Technique Using Opencv and Python

YCbCr (Luminance Chrominance value): YCbCr consists of three color channels in brightness and a two-color channel. Y for brightness channel (luma), Cb is a blue—luma (B-Y) and Cr represents red—luma (R-Y) channel (Fig. 3) [14].

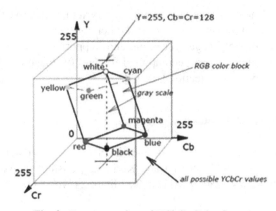

Fig. 3. Representation of YCbCr Color Space

2.2 Classification Using Sklearn and Python

The supervised machine learning of SVM algorithm have been used for classifying EUS disease fish area and normal fish area. SVM takings a image and plot with class to predict which EUS disease fish area or normal fish area presents in a selected image. The kernel RBF (Radial Basis Function (RBF) is defined in Eq. (1) to plot the class for this study

$$K(X, X') - \exp\left(-\frac{\|X - X'\|}{2\sigma^2}\right) \tag{1}$$

Euclidean distance is squared $\|x - x'\|^2$ of two data points of an image x and x'. The decision boundary is drawn between EUS disease fish area class and normal fish area class by generating hyperplane using this algorithm. The hyperplane is used to detect the input image class [15, 17].

2.2.1 SVM Performance Evaluation

The metrics technique has been used for performance to detect EUS disease fish area and normal fish area. These are values of Accuracy, Precision, Recall and F1-Score described in Eq. (2)–Eq. (5).

$$\text{Accuracy} = (tp + tn)/(tp + fp + fn + tn) \tag{2}$$

$$\text{Precision} = (tp)/(tp + fp) \tag{3}$$

$$\text{Recall} = (tp)/(tp + fn) \tag{4}$$

$$\text{F1-Score} = (2tp)/(2tp + fp + fn) \tag{5}$$

To calculate the number of truly classified positive sample pixels in the detected area is using True Positive (tp). To calculate the number of truly classified negative sample pixels in the detected area is using True Negative (tn). False Negative (fn) is applied to calculate the number incorrectly classified negative sample pixels in detected area. False Positive (fp) is applied to calculate the number incorrectly classified positive sample pixels in detected area [18].

2.2.2 Experimentation SVM with Test Dataset

Two types of test dataset use for SVM experimentation these are

i) Colour-image consist of 66 nos of EUS disease fish area Vs 66 nos of normal fish area and
ii) Cr-image consist of Cr-Channel Image of YCbCr 66 nos of EUS disease fish area Vs Cr-Channel Image of YCbCr 66 nos of normal fish area

3   Results and Discussion

3.1   YCbCr and Segmentation

YCbCr Convert RGB Image than YCbCr toY-Channel Image, YCbCr to Cr-Channel Image ,YCbCr to Cb-Channel (Fig.4)

 i) EUS disease fish area of image sample (Fig. 5)
ii) Normal fish area of image sample (Fig. 6)

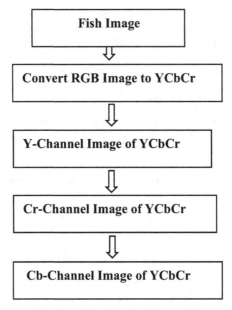

Fig. 4. Complete process of YCbCr Y-Channel

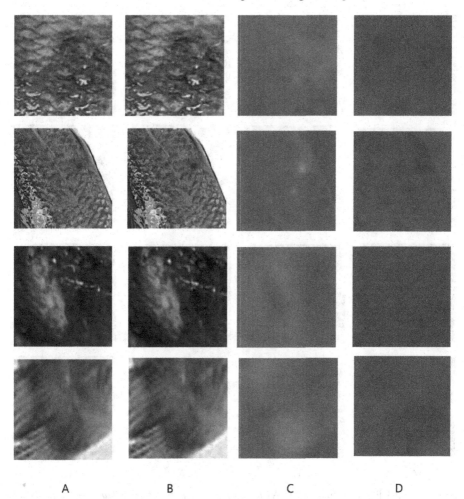

Fig. 5. EUS disease fish area images of (A) *Labeo calbasu, Labeo bata, Channa striata, Mystus tengara* (B) Y-Channel of YCbCr (C) Cr-Channel of YCbCr (D) Cb-Channel of YCbCr

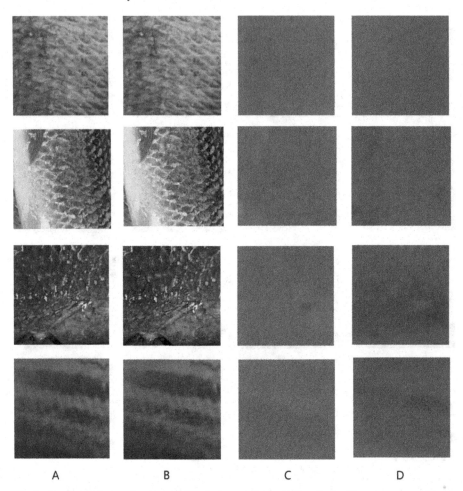

A   B   C   D

Fig. 6. Normal fish area images of (A) *Labeo calbasu, Labeo bata, Channa striata, Mystus tengara* (B) Y-Channel of YCbCr (C) Cr-Channel of YCbCr (D) Cb-Channel of YCbCr

The result YCbCr three colour channels Y, Cb and Cr of infected with EUS disease fish area as well as normal fish area *Labeo calbasu, Labeo bata, Channa striata , Mystus tengara* are shown in Fig 5 and Fig 6. It is clearly visible some of the disease area Y channel and Cr channel because Cr channel represent the red colour component of the image. Similarly Cr channel in YCbCr colour space is used in skin with wound detection from image [19].

3.2 SVM Performance

The performance RBF kernel based SVM is processed on Colour-image and Cr-image dataset on Table 1.

Table1 shows a comparison among the investigations mentioned. Test dataset Colour-image achieves 75% accuracy, 75% precision, 83% recall and 73% F1-score and test dataset Cr-image achieves 85% accuracy, 85% precision, 88% recall and 84% F1-score,which are competitively 10% accuracy, 10% precision, 5% recall and 9% F1-higher values Cr-image than the Colour-image and confusion matrix (Fig 7). Similar SVM performance used in Salmon infected fish and skin cancer detection from images.

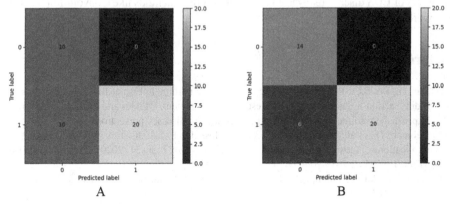

Fig. 7. (A) Confusion matrix Colour-image, (B) Confusion matrix Cr-image

Table 1. SVM Dataset wise performance

Type of dataset	Class	Accuracy	Precision	Recall	F1-Score
Colour-image	disease	**0.75**	0.50	1.00	0.67
	normal	**0.75**	1.00	0.67	0.80
	avg	**0.75**	0.75	0.83	0.73
Cr-image	disease	**0.85**	0.70	1.00	0.82
	normal	**0.85**	1.00	0.77	0.87
	avg	**0.85**	0.85	0.88	0.85

4 Conclusion

The early detection fish disease can increase the fish production. The present study show the dominant colour EUS diseased of fish area clearly using Cr segmentation.

The proposed method for Support Vector Machine (SVM) of machine learning technique for detecting EUS disease fish area successfully above 75% to 85%.

In future to achieve 100% accuracy detection need large datasets and increase the numbers of EUS disease fish varieties as well as other methods like Convolution Neural Network, GNN etc.

References

1. Kar, D., De, S.C., Roy, A.: Epizootic Ulcerative Syndrome in Fish at Barak Valley, Assam, India. Section-8 Sustainable Water Resource Management, Policies and Protocols (2000)
2. https://fisheriesdirector.assam.gov.in/portlets/aquatic-diseases-preventions
3. Debnath, D., et al.: Assessment of economic loss due to fish diseases in Assam, India and implications of farming practices. J. Inland Fisheries Soc. India **51** (2020). https://doi.org/10.47780/JIFSI.51.2.2019.106499
4. Kalita, B., Ali, A., Islam, S., Hussain, I., Pokhrel, H.: Incidence of fish diseases in Assam, pp. 814–817 (2019)
5. Chakravorty, H., Paul, R., Das, P.: Image processing technique to detect fish disease. Int. J. Comput. Sci. Secur. (IJCSS) **9**(2), 121–131 (2015)
6. Chakravorty, H., Das, P.: Simulink model based fish disease analysis. Int. J. Inf. Sci. Comput. **2**(1), 121–131 (2015)
7. Chakravorty, H.: To detection of fish disease using augmented reality and image processing. Eur. J. Appl. Sci. **7**(6), 01–04 (2020). https://journals.scholarpublishing.org/index.php/AIVP/article/view/7503
8. Chakravorty, H.: New approach for disease fish identification using augmented reality and image processing technique. IPASJ Int. J. Comput. Sci. (IIJCS) **9**(3), 1–7 (2021)
9. Rachman, F., Akbar, M., Putera, E.: Fish disease detection of epizootic ulcerative syndrome using deep learning image processing technique. In: Proceedings International Conference on Fisheries and Aquaculture, vol. 8, no. 1, pp. 23–34 (2023). https://doi.org/10.17501/23861282.2023.8102
10. John, N., Viswanath, A., Vishvanathan, S., Kp, S.: Analysis of various color space models on effective single image super resolution **384**, 529–540 (2016). https://doi.org/10.1007/978-3-319-23036-8_46
11. Mustafa Abdullah, D., Mohsin Abdulazeez, A.: Machine learning applications based on SVM classification a review. Qubahan Acad. J. **1**(2), 81–90 (2021). https://doi.org/10.48161/qaj.v1n2a50
12. https://latitude.to/articles-by-country/in/india/73836/hailakandi
13. https://latitude.to/articles-by-country/in/india/38333/karimganj
14. https://pub.towardsai.net/opencv-different-color-spaces-in-image-processing-with-python-17bbed3592ad
15. https://medium.com/analytics-vidhya/image-classification-using-machine-learning-support-vector-machine-svm-dc7a0ec92e01
16. https://github.com/aditi-govindu/Image-Classsification-using-sklearn/blob/main/Image_Classification_using_SVM.ipynb

17. Ahmed, M.S., Aurpa, T.T., Azad, M.A.K.: Fish disease detection using image based machine learning technique in aquaculture. J. King Saud Univ. Comput. Inf. Sci. **34**(8), 5170–5182 (2022). https://doi.org/10.1016/j.jksuci.2021.05.003. (https://www.sciencedirect.com/science/article/pii/S1319157821001063). Part A. ISSN 1319-1578
18. Babu, G.N.K., Joseph Peter, V.: Skin cancer detection using support vector machine with histogram of oriented gradients features. ICTACT J. Soft Comput. **11**(02) (2021)
19. Li, F., Wang, C., Xiaohui, L., Peng, Y., Jin, S.: A composite model of wound segmentation based on traditional methods and deep neural networks. Comput. Intell. Neurosci. **2018**, 1–12 (2018). https://doi.org/10.1155/2018/4149103

Author Index

Printed in the United States
by Baker & Taylor Publisher Services